VOL.17

7817

JOURNAL OF ARMORED

ASSAULT

& HELIBORNE WARFARE

603-609 Castle Peak Road
Kong Nam Industrial Building
10/F, B1, Tsuen Wan
New Territories, Hong Kong
www.concord-publications.com

We welcome authors who can help expand our range of books. If you would like to submit material, please feel free to contact us.

We are always on the look-out for new, unpublished photos for this series. If you have photos or slides or information you feel may be useful to future volumes, please send them to us for possible future publication. Full photo credits will be given upon publication.

ISBN 962-361-132-3
printed in Hong Kong

Offspring of the Weasel Family

The New Aufklärungsfahrzeug Wiesel 1

Ralph Zwilling

This recon section of Luftlandeaufklärungskompanie 310 is taking part in a hostage rescue operation, which is part of the training rotation of the DSO at the German Army Combat Training Center. Note the tied-down radio antennas, as well as the storage net stowed on the front of the vehicle. Because of the cold weather, all soldiers wear winter uniforms and scarves.

Preliminary considerations to provide a new mission-specific vehicle to Luftlandeaufklärungskompanie 310 (Airborne Reconnaissance Company 310) in Lüneburg, Germany, started in 1999. This unit was scheduled for activation in 2000, but from the beginning the German Bundeswehr planned to procure only sixteen vehicles for it. According to military, tactical and technical factors, the newly developed air-transportable, armored, off-road vehicle was intended to provide its crew with capabilities in conducting battlefield reconnaissance missions behind enemy lines. This would be done over long distances from their own forward lines whether by night or day.

Very early in the development process, the Bundeswehr decided to use the Wiesel 1 Waffenträger (Weapon Carrier) as the basis for this new airborne reconnaissance vehicle, as it had already gained considerable experience with this reliable family of vehicles. From 1990 to 1992, German Army Airborne Troops had received 133 Wiesel 1 MK 20, and 210 Wiesel 1 TOW vehicles, to replace the ageing Kraftkarren (KRAKA), which was withdrawn from Airborne Brigade service in 1989. The UNOSOM mission in Somalia in 1993 was the first mission in which the Wiesel 1 was involved. Since that time the Waffenträger Wiesel 1 has proven to be very reliable during operations in foreign countries like Bosnia, Kosovo and Macedonia. It is also serving in Afghanistan, where the vehicles are used for convoy escort and security operations. Furthermore, they are used to control Areas of Responsibility (AORs) and other objectives, to observe and secure checkpoints, as well as to support troops as they conduct security missions.

The newer Wiesel 2 versions are also very popular amongst the soldiers of the Bundeswehr. Compact dimensions and proven air transportability by the CH-53G Sikorsky medium cargo helicopter, and cargo airplanes like the C-160 Transall and the C-130 Hercules, were all determining factors in the selection of the Wiesel 1 as the foundation for a newly developed

The new armored, air-transportable and off-road Aufklärungsfahrzeug Wiesel 1. It is produced by Rheinmetall Landsysteme GmbH, and provides its three-man crew with the capability to conduct reconnaissance missions on the battlefield over long distances behind enemy lines by night and day.

The new AufklFzg Wiesel 1 was created by converting and modifying existing Wiesel 1 Waffenträger vehicles that were fitted with the TOW ATGM (Anti-Tank Guided Missile). Out of ergonomic considerations, the roof of the rear crew compartment was raised by a number of centimeters.

airborne reconnaissance vehicle. Rheinmetall Landsysteme GmbH (RLS) delivered a pre-series Aufklärungsfahrzeug Wiesel 1 (AufklFzg Wiesel 1 / Reconnaissance Vehicle Wiesel 1) to the German Army's Wehrtechnische Dienststelleas in July 2001 for evaluations and troop trials. This vehicle was subsequently upgraded to the later series level. During its development, the prototype vehicle was subjected to a number of tests. These included shelling, bombing, being dropped from a crane, fatigue strength tests, corrosion tests, ergonomic tests, and numerous driving course evaluations both on and off the road. After the successful completion of this extensive test program, sixteen vehicles were modified and delivered to the Bundeswehr between January and March of 2005. They were then handed over to Luftlandeaufklärungskompanie 310.

To achieve the alterations to the basic upper hull, extensive conversions and adaptations had to be conducted by Rheinmetall Landsysteme GmbH engineers and workers. Amongst other modifications, the entire TOW ATGM weapon station was removed, the old crew seats were replaced by new ergonomically designed seats, and an advanced reconnaissance and radio system was installed in the small, narrow vehicle.

The 80-liter fuel tank, located outside the armored hull at the rear of the vehicle, provides the vehicle with a very respectable range of 200-300km, depending on the terrain. Note the white Tarnkreuz below the fuel tank, which is used for convoy operations at night or during operations in low visibility conditions.

Combined battlefield operations. This recon section was seen during a rotation of the Division Spezielle Operationen (DSO) at the Gefechtsübungszentrum Heer (Army Combat Training Center) in January 2006. Because of a very low ground pressure of only 3.6N/cm², the Aufklärungsfahrzeug Wiesel 1 still remains mobile even when other combat vehicles can no longer operate effectively.

The new AufklFzg Wiesel 1 was created by converting and modifying the existing Wiesel 1 Waffenträger armed with the TOW Anti-Tank Guided Missile (ATGM). Because of ergonomic guidelines, the roof of the rear crew compartment was raised by a number of centimeters. To achieve this basic alteration of the upper hull, extensive conversions and adaptations had to be conducted by Rheinmetall Landsysteme GmbH engineers and workers. Amongst other modifications, the entire TOW ATGM weapon station was removed, the old crew seats were replaced with new ergonomically designed seats, and advanced reconnaissance and radio systems were installed in the small, narrow vehicle.

Luftlandeaufklärungskompanie 310 was activated on 1 April 2000 in Lüneburg, Germany, and was attached to Panzeraufklärungsbataillon 3 (Reconnaissance Battalion 3) of 1. Panzerdivision. Starting on 1 January 2006, it was directly attached to Luftlandebrigade 31 in Oldenburg, Germany. This latter unit belongs to the Division Spezielle Operationen (DSO – Division Special Operations). It is planned that the unit will consist of a company headquarters section, a supply section, a signals section, a maintenance team for the Unmanned Aerial Vehicle (UAV) KZO, a technical section, a transportation section, two medical teams, and two reconnaissance platoons that are each equipped with six AufklFzg Wiesel 1 vehicles.

The reconnaissance platoon will be organized into three recon sections, each with two AufklFzg Wiesel 1 vehicles. The remaining four vehicles will be held in reserve for operations. Furthermore, the company will have a technical reconnaissance platoon with a platoon headquarters, plus three radar sections, each of which has a LEGAR (leichtes Gefechtsfeldaufklärungsradar / Light Battlefield Surveillance Radar) and six sensor teams. To support the AufklFzg Wiesel 1 recon sections, which have a maximum weight and capacity limit for air transportability, each sensor team will be equipped with a ground sensor unit and an Aladin UAV. The LEGAR provides central handling of all functions associated with integrated battlefield reconnaissance. From computer-based mission planning, automatic detection, tracking and identification, right up to sending target grid codes into a command and control system, all the necessary functions for a flexible and autonomous operation are integrated. Consisting of the BOR-A 550 radar, as well as one HF and two VHF radio systems, the LEGAR is carried on a long-wheelbase Mercedes Benz Lkw 0.9t gl leicht Wolf. Depending on mission requirements, the entire system can be controlled from a distance of up to 100m from the vehicle. An autonomous power supply enables a 72-hour operation separate from the vehicle.

It is envisioned that Luftlandeaufklärungskompanie 310 in Lüneburg, Germany, will have two reconnaissance platoons that are each equipped with six Aufklärungsfahrzeuge Wiesel 1 vehicles. The reconnaissance platoon will be organized into three recon sections, each with two AufklFzg Wiesel 1 vehicles. Note the shock absorber between the first and second road wheels.

The high mobility and outstanding performance of the Wiesel Family of Vehicles (FOV) in general and of the Aufklärungsfahrzeug Wiesel 1 in particular, results in speed and agility on the one hand, and relatively easy and quick air transportability on the other. The former Soviet Army Barracks in Hillersleben, Germany, are perfect for Military Operations in Urban Terrain (MOUT) training exercises.

A low weight, a low ground pressure, and diminutive dimensions, enable the transport of two fully equipped and armed vehicles inside a CH-53G Sikorsky medium cargo helicopter. If necessary, the tank can also be transported sling-loaded underneath. Note the tied-down radio antenna and the storage box attached to the front of the vehicle.

The engine pack located in the front left of the hull includes the engine, the air filter system, the cooling system, the exhaust system, the automatic transmission and the steering transmission. Training Rotation 01/2006 was the first exercise that the Aufklärungsfahrzeuge Wiesel 1 was involved in at the Army Combat Training Center in Letzlingen, Germany.

The new vehicle of the DSO cannot swim, but it can easily ford water obstacles up to 50cm deep. With its compact dimensions (a width of 1.82m and a length of 3.55m), it is easy for the Wiesel 1 to negotiate the narrow streets of Afghanistan and Kosovo. Note the personal equipment attached to the rear of the AufklFzg Wiesel 1, and the antenna rods.

If necessary, the commander can also use a Leica handheld telescope with a 20-60x magnification, which can be optionally mounted on a steel rail located in front of his hatch. When not in use, this telescope is fixed with straps to the roof between the commander's and scout's seats. Unlike the new Spähwagen Fennek, and because of weight and dimension limitations, the sensor head of the Aufklärungsfahrzeug Wiesel 1 cannot be tripod-mounted at a separate location for remote operation.

The military intelligence platoon will be organized into a platoon headquarters, an evaluation team, and five military intelligence teams slated to be equipped with the new Einsatzfahrzeug Spezialisierte Kräfte (ESK) MUNGO. Even though the Unmanned Aerial Vehicle (UAV) platoon will have six Kleinfluggeräte für Zielortung (KZO) UAVs, only one system can be used at a time. The KZO uses an infrared sensor system to acquire real-time images of targets, as well as the position and speed of either moving or stationary targets. The target data can be transmitted from the air vehicle to the ground station over a distance of more than 100km, allowing the target data to be effectively used by other assets such as long-range artillery, intelligent artillery munitions, rockets and missiles. The air vehicle system can be deployed in all weather conditions by day or night.

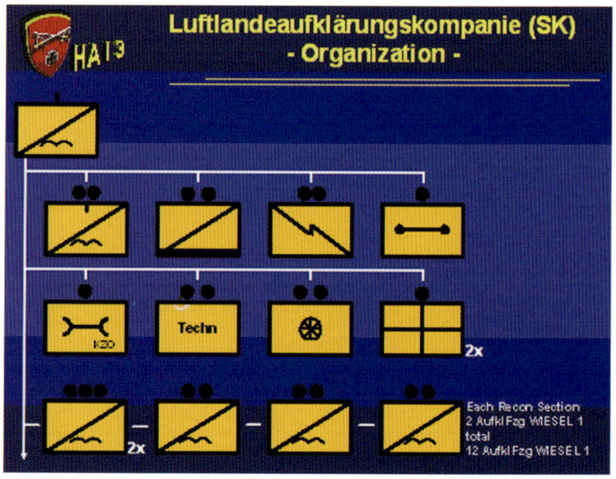

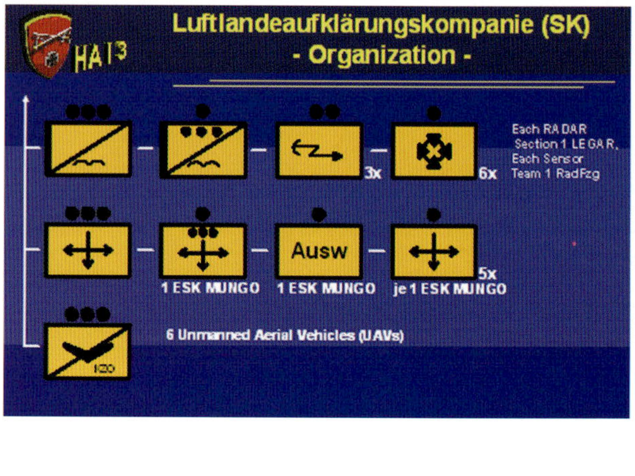

This recon section consisting of two Aufklärungsfahrzeuge Wiesel 1 vehicles conducted an area reconnaissance at the Hillerleben MOUT site to gain as much information as possible for the trailing main force. The training of LLAufklKp 310 is oriented towards the respective mission capabilities required of the airborne reconnaissance and technical reconnaissance sections and teams in the DSO.

Equipped with the latest small-arms weapons, powerful long-range radios, and advanced personal equipment, the soldiers of Luftlandeaufklärungskompanie 310 are able to operate and provide reconnaissance information in all climatic zones for the Division Spezielle Operationen (DSO). Even though the German Bundeswehr still has many conscripts that serve for only nine months, most of the company's personnel are professionals or soldiers that are serving for a fixed period. This concept ensures that the soldiers are available for a long period after finishing their extensive prerequisite training. The specific mission of LLAufklKp 310 within special operations, and the challenging training, requires a careful selection of personnel. The training of LLAufklKp 310 is oriented towards the required capabilities needed by the airborne reconnaissance sections and the technical reconnaissance sections and teams in their possible mission options within the Division Spezielle Operationen. These special capabilities are reconnaissance and reconnaissance patrols, monitoring of areas, detection of weapons caches, radar detection, ensuring a command and control capability in the entire operational area (AOR), and the capacity to deploy immediately after notification via helicopters, cargo aircraft or parachutes. In the near future, Luftlandeaufklärungskompanie 310 will very likely be moved from Lüneburg to Seedorf to support Luftlandebrigde 31 "Oldenburg" (Airborne Brigade) more effectively with reconnaissance information.

The Aufklärungsfahrzeug Wiesel 1 in detail

The high mobility and outstanding performance of the Wiesel Family of Vehicles (FOV) in general and of the Aufklärungsfahrzeug Wiesel 1 in particular, results in speed and agility on the one hand, and relatively quick and easy air transportability on the other. Its low weight, low ground pressure and compact dimensions enable the transport of two fully equipped and armed vehicles inside a CH-53G Sikorsky medium cargo helicopter. If necessary, the tank can also be transported sling-loaded underneath. At the moment the German Bundeswehr is conducting tests to air drop the AufklFzg Wiesel 1 using special, cushioned shock-absorbing pallets, but up till now there have not been satisfactory results. Furthermore, the vehicle can be transported by either truck or ship to the AOR. Four AuflFzg Wiesel 1 vehicles, their three-man crews, and all their armaments and equipment, can be transported by a single C-160 Transall, or by two C-130 Hercules cargo aircraft. To safely tie down the vehicle during transportation, it is equipped with twelve mounting eyes located all around the hull. Strategically important regions, which cannot be reached because of an undeveloped infrastructure or a lack of paved streets, are not

By using the telescoping mast-mounted sensor head (which can be raised up to its highest position 220cm above the ground), the crew has the possibility of taking advantage of effective natural or artificial cover while observing the battlefield by day or night from concealed positions.

Because of the advanced and proven torsion bar suspension consisting of return rollers, hydraulic shock absorbers and ebonite bump stops, the new vehicle has outstanding off-road mobility. As with the Wiesel 1 Weapon Carrier, the braking system of the new Aufklärungsfahrzeug Wiesel 1 also consists of a hydraulic dual-circuit disc brake system, with a service brake and a mechanical parking brake.

Strategically important regions that cannot be reached because of a lack of infrastructure or paved streets are not an insurmountable problem for the new Aufklärungsfahrzeug Wiesel 1. This ability qualifies the Wiesel FOV to serve as the eyes and ears of the airborne troops of the DSO.

From the beginning, the German Bundeswehr planned to procure only sixteen vehicles. Very early in the development process, the Bundeswehr decided to use the Wiesel 1 Waffenträger (Weapon Carrier) as the basis for its new airborne reconnaissance vehicle, as it had already gained very good experience with this reliable family of vehicles.

Immediately before the main elements of Fallschirmjägerbataillon 263 (from Zweibrücken, Germany) entered the MOUT site at Hillersleben in January 2006, this Aufklärungsfahrzeug Wiesel 1 conducted a recon mission to get more information about enemy elements and reconnoiter the best route to the hospital and community center.

an insurmountable problem for the new Aufklärungsfahrzeug Wiesel 1. These abilities have predestined the Wiesel FOV to this new application with the DSO, where they are to be used as the eyes and ears of the airborne troops.

Above all, this new vehicle of the DSO is characterized by outstanding performance and high mobility during on-road and off-road operations. Quick acceleration guarantees a swift change of fighting positions, which also increases its survivability on the battlefield. The engine pack is located in the left front of the hull – the pack includes the engine, the air filter system, the cooling system, the exhaust system, the automatic transmission and the steering transmission. The water-cooled 64kW 5-cylinder Volkswagen diesel engine with turbocharger (it is derived from a civilian engine) provides the 3.23-ton tank with a maximum speed of 80km/h on roads and 60km/h off-road. A well-trained maintenance team only needs fifteen minutes to change the entire engine pack. Both the 3-gear 3HP-22 automatic transmission and the down-streamed 2-gear steering transmission are produced by the well-known company of Zahnradfabrik Friedrichshafen (ZF). Hydraulic steering brakes in a Cletac steering transmission are used to steer the vehicle and provide a minimum turning radius of only 5.25m. Like the Wiesel 1 Weapon Carrier, the braking system of the new Aufklärungsfahrzeug Wiesel 1 also consists of a hydraulically operated dual-circuit disc brake system that offers a service brake plus a mechanical parking brake.

The 80-liter fuel tank, located outside the armored hull at the rear of the vehicle, provides the vehicle with a very respectable range of 200-300km, depending on the terrain. Because of the advanced and successful torsion bar suspension that incorporates return rollers, hydraulic shock absorbers and ebonite bump stops, the new vehicle has an outstanding off-road mobility. Initially, the Wiesel 1 Weapon Carrier was equipped with continuous rubber tracks, but these were later replaced by the new Type

Readers can best imagine the small size of the new Aufklärungsfahrzeug Wiesel 1 if it is compared with the huge Leopard 2A4 Main Battle Tank. When the paratroopers of Fallschirmjägerbataillon 263 entered Hillersleben, the Leopard 2A4s provided cover.

Compact dimensions, and proven air transportability by CH-53G Sikorsky medium cargo helicopters and cargo aircraft like the C-160 Transall and the C-130 Hercules, were determining factors in choosing the Wiesel 1 as the basis for this newly developed airborne reconnaissance vehicle. Note the mesh covering the exhaust system and the small "Eiserne Kreuz" tactical marking on the side of the vehicle.

Before the paratroopers of the Division Spezielle Operationen started a hostage rescue operation at the northern boundary of the German Army Training Center in Letzlingen, two Aufklärungsfahrzeuge Wiesel 1 vehicles of Luftlandeaufklärungskompanie 310 conducted a reconnaissance mission to get more information about the number of enemy soldiers, and how the hostage rescue team could best infiltrate the MOUT site.

622 C steel track with replaceable rubber pads. This track is produced by Diehl Remscheid GmbH & Co. KG, and is of the same type used on the AufklFzg Wiesel 1. This new track provides greater reliability and requires less maintenance. Because of a very low ground pressure of only 3.6N/cm?, the Aufklärungsfahrzeug Wiesel 1 is still mobile even when other combat vehicles can no longer operate effectively. Agilely, and without too much effort, it can handle 60% gradients and side slopes at angles of up to 30%. In addition, trenches up to 120cm wide and obstacles up to 40cm high pose no problem. The new vehicle of the DSO cannot swim, but it can easily ford water obstacles up to a depth of 50cm. With its compact dimensions (a width of 1.82m and a length of 3.55m), it is easy to operate in the narrow streets of Afghanistan and Kosovo.

In keeping with the Wiesel 1 Weapon Carrier, the driver's station is located in the right front area of the hull. The hull itself consists of steel plates of varying thickness and consistency, which provide the three-man crew sufficient protection against shrapnel produced by different weapons such as mines or artillery grenades, as well as from infantry ammunition. The compact instrument panel with various displays (including a speedometer, rev counter, battery power display and fuel display) is located on the right side of the driver's seat. To operate the small tank while sitting beneath the closed armored hatch, the driver has three periscopes. He can easily replace the center periscope with a night vision device for nighttime driving. The stations of the commander and the scout are parallel to each other immediately behind the driver.

Specially designed seats from the German company of Autoflug GmbH were installed to reduce weight and to provide a comfortable seating position. These seats were also utilized during the upgrade of the Leopard 2A6M Main Battle Tank. Each of the seats, which are made from a very hardwearing green canvas material, is attached to the vehicle's roof with four straps so that they have no direct contact with the floor plate. To provide the crew with an ergonomically comfortable position while observing and scanning the surrounding area through the open hatches, these mine-protection seats can be locked in different positions by a sophisticated but very easy to use adjustment system attached to the roof.

Besides three 5.56mm x 45 G36A1 assault rifles, the crew can also field a 7.62mm x 51 MG3A1 machine gun, and a 40mm grenade pistol for self-defense. The 3.77kg G36A1 is constructed almost entirely of tough, glass fiber reinforced polymer material. The main weapon components consist of glass fiber reinforced polymer with stainless steel inserts. Thanks to the self-regulating gas block, the gas-operated weapon works with extreme reliability and only requires a minimum of maintenance.

The dependable radio system consists of a SEM 80/90 (VHF), a HRM 7400 (HF), and the handheld HRS 7000 Manpack (HF) and SEM 52SL (VHF) systems. Due to a lack of space, the ATU 7400 antenna tuning unit is located outside the crew compartment on the rear of the hull.

By pushing a switch, the soldiers can drop back into the safety of the crew compartment within seconds.

As the interior of the Aufklärungsfahrzeug Wiesel 1 is very small and narrow, nearly every area is used to store equipment. On the wall of the rear compartment, to the right of the commander's seat, is a 40mm grenade pistol from Heckler & Koch, a storage bag, and the mount for a 5.56mm x 45 G36A1 rifle. Right behind the seat, the advanced hydride navigation system LLN-GY, made by LITEF, is located. It consists of an LLN-GY inertia unit, a PLRG96 GPS receiver, a GPS antenna, and an EWG 20 odometer that permits accurate determination of the vehicle's position and direction with respect to grid north. In front of the scout's seat there is a mount for another night vision device. Both the scout's and the driver's 5.56mm x 45 G36A1 rifles, plus a first aid kit and another small storage bag, are attached to the wall on his left. The area behind the scout's seat is used to store spare barrels for the machine gun, and parts for the extensive VHF and HF radio system.

The heart of the Aufklärungsfahrzeug Wiesel 1, costing nearly ?1.4 million, is the Autonomes Optronisches Zielsystem (AOZ) 2000. This technically advanced and proven observation and reconnaissance unit is produced by Rheinmetall Defence Elektronics GmbH. This system consists of a telescoping mast-mounted sensor head (Schwenkneigekopf – SNK) that includes an OPHELIOS thermal

This commander is seen scanning the surroundings with his 20-60x Leica telescope attached to a steel rail in front of his hatch. Equipped with the latest small arms, powerful long-range radios, and advanced personal equipment, the soldiers of Luftlandeaufklärungskompanie 310 are able to operate and provide reconnaissance information in all climatic zones for the DSO.

To operate the small tank while sitting below the closed armored hatch, the driver has three periscopes. He can easily replace the central periscope with a night vision device. The stations of the commander and the scout are parallel to each other right behind the driver. Note the covered license plate and headlights.

Skillfully and without too much effort, the vehicle handles 60% gradients and up to 30% side slopes. Trenches up to a width of 120cm, and obstacles up to a height of 40cm, are also no problem. The new vehicle of the DSO cannot swim, but it easily fords water obstacles up to 50cm deep.

With the scouts ready to provide machine gun fire support if necessary, the commanders of this recon section scan the main street of Hillersleben for enemy activity. Starting in 2006, the German Army Combat Training Center will conduct twenty-one training rotations per year.

imager, a CCD day vision camera, a HALEM 2 laser rangefinder, and a sight system control unit. During missions, the commander normally observes the battlefield with the compact sensor head. Depending on mission requirements, he is supported by both the scout and the driver. Within just six seconds, the sensor head can be raised up to its highest position 220cm above the ground. Thereby the crew can take effective advantage of natural or artificial cover while observing the battlefield by day or night from their concealed position. If necessary, they can also use a Leica handheld telescope with a 20-60x magnification, which can be optionally mounted to a steel rail located in front of the commander's hatch. When not in use, this telescope is fixed with straps to the roof between the commander's and the scout's seats.

Unlike the new Spähwagen Fennek, and because of weight and dimension limitations, the sensor head of the Aufklärungsfahrzeug Wiesel 1 cannot be tripod-mounted at an exposed location for remote operations. However, vehicle or personnel targets can be detected out to distances of 2,500 meters with the help of the OPHELIOS thermal imager, even at night or during times of limited visibility. Definite target identification can be achieved from a distance under 1,800 meters. The compact black and white sight display system mounted below the roof on the right side of the commander's seat (which can also be swiveled towards the scout if necessary) provides the operator with the ability to easily evaluate the reconnaissance images of both the thermal imager and the CCD day vision camera. When switching from the thermal imager, with two observation magnifications, to the CCD day vision camera, the observation view is automatically adopted, this helping to keep track of a target during the transformation from thermal to day sight. Using the small, black joystick of the sensor head control unit that is located to the right to the commander's station, the sensor head can be easily controlled 360° in azimuth and -8° to +60° in elevation.

Determination of target coordinates is achieved using the system's laser rangefinder and azimuth and elevation measuring equipment, together with the navigation system. To locate a target, the operator just has to push a red button on the control unit joystick. He can choose between the first and last laser echo. Depending on the weather conditions and the target properties, the laser system can be used at any distance between 50 and 5,000 meters, with a margin for error of only about fifty meters. Then the operator inputs the target information into the new Führungs- Kommunikations- und Informationssystem (FaKom - Command, Control and Communication System) built by Rheinmetall Defence Electronics, which is located in the small space between the commander's and the scout's stations. This very advanced C^3 system consists of a touch-screen display and control unit, a computer module behind the commander's seat, the data terminal MIDAT with a memory card, and a keyboard. For orientation in cross-country terrain, digitized maps that show the vehicle's position, target position, and current battlefield environment, are used. Information is easily displayed as an icon on the FaKoM touch screen and is continuously updated between vehicles and the command post via HF and VHF radio data links. The AOZ 2000 control unit located below the FaKoM includes controls for brightness, contrast, and focus, as well as the magnification and focus switches for the thermal imager and the CCD day vision camera.

The water-cooled 64kW five-cylinder Volkswagen diesel engine with turbocharger (which originates from the civilian market) provides the 3.23-ton tank with a maximum speed of up to 80km/h on roads and 60km/h off-road. A well-trained maintenance team only needs fifteen minutes to change the entire engine pack.

Reconnaissance is teamwork! Even though the German Bundeswehr still has many conscripts serving for only nine months, most of the company's soldiers are professional soldiers, or else soldiers serving for a fixed period. This concept ensures that the soldiers are available for a long period after finishing their extensive training. The crew of the new Aufklärungsfahrzeug Wiesel 1 produced by Rheinmetall Landsysteme GmbH consists of the commander, the scout and the driver.

The reliable 7.62mm x 51 MG3A1 machine gun is operated by the scout and attached to a pivot mount in front of his hatch. With the MG3A1, 120 ready-to-fire 7.62mm rounds are available, while another five ammo boxes (each with 120 rounds) are stored in the crew compartment. Operation and reloading of the machine gun cannot be conducted from under armor protection, so the scout has to stand in the open hatch.

While elements of Fallschirmjägerbataillon 263 enter the Hillersleben MOUT site, this commander scans the hospital and the marketplace, which were expected to be a stronghold of the opposing forces. Note the mesh wire material covering the lens of the Leica telescope, which is designed to prevent reflection of sunlight.

Some parts of the extensive radio system are located in the space between the engine compartment and the FaKoM. The reliable radio system consists of an SEM 80/90 (VHF), a HRM 7400 (HF), and the HRS 7000 Manpack (HF) and SEM 52SL (VHF) handheld systems. Because of the lack of space, the ATU 7400 antenna tuning unit is located outside the crew compartment on the rear of the hull. Often another portable SEM 70 (VHF) radio, and two 360g Lucie night vision goggles (produced by Thales Defence Deutschland GmbH), are stored in the gap between the commander's and the scout's seats.

For self-defense, the crew can also field a 7.62mm x 51 MG3A1 machine gun in addition to the aforementioned 40mm grenade pistol and three 5.56mm x 45 G36A1 rifles. The reliable machine gun is operated by the scout and attached to a pivot mount in front of his hatch. With the MG3A1, a total of 120 7.62mm ready-to-fire rounds are available, and another five ammo boxes each with 120 rounds are stored in the crew compartment. Operation and reloading of the machine gun cannot be carried out from under armor protection, so the scout has to stand in the open hatch. Because of weight and capacity limitations, and in absolute contrast to the recently introduced Spähwagen (Reconnaissance Vehicle) Fennek, the Aufklärungsfahrzeug Wiesel 1 has no smoke discharger system to create a smokescreen approximately fifty meters in front of the vehicle. The AufklFzg Wiesel 1 also lacks a rear view camera, a fire extinguishing system, or a nuclear, biological and chemical (NBC) warfare protection system. The only luxury provided to the crew is the small auxiliary heating system. Because of the limited space inside the narrow crew compartment, some personal equipment like the three sleeping bags, water and numerous MREs (Meals, Ready to Eat) have to be stored in the large metal box attached to the front of the vehicle between the front lights.

To enlarge the crew compartment, the roof of the old Wiesel 1 TOW Waffenträger was raised by a number of centimeters. Note the HF antenna mount and the crew's NBC bags attached to the side of the Aufklärungsfahrzeug Wiesel 1. The scout provides fire support with his G36A1 assault rifle while the commander provides information to the task force headquarters via the advanced radio system.

The future will show if Rheinmetall Landsysteme GmbH will get further orders from the German Bundeswehr or other armies to produce more Aufklärungsfahrzeuge Wiesel 1 vehicles. Nevertheless, this new reconnaissance vehicle greatly enriches the Wiesel Family of Vehicles and it will provide good service to the soldiers of the Division Spezielle Operationen. It is just a matter of time before the first Aufklärungsfahrzeug Wiesel 1 will be seen in Afghanistan in use with Germany's GECON ISAF troops.

The author would like to thank Oberleutnant Bammel (GefÜbZ H), Hauptmann Paare (LLAufklKp 310), Stabshauptmann Assenbrunner (Heeresamt), Oberstleutnant i.G. Broich (Heeresamt), Oberstleutnant i.G. Schmidt (Bundesministerium der Verteidigung), and the soldiers of Luftlandeaufklärungskompanie 310 in Lüneburg for their outstanding support.

The heart of the Aufklärungsfahrzeug Wiesel 1 is the technically advanced and proven observation and reconnaissance unit known as Autonomes Optronisches Zielsystem (AOZ) 2000, produced by Rheinmetall Defence Elektronics GmbH. This system consists of a telescoping mast-mounted sensor head (Schwenkneigekopf - SNK) that includes an OPHELIOS thermal imager, a CCD day vision camera, a HALEM 2 laser rangefinder, plus a sight system control unit.

The telescoping mast-mounted sensor head (Schwenkneigekopf - SNK) in detail. It includes an OPHELIOS thermal imager, a CCD day vision camera, a HALEM 2 laser rangefinder, and a sight system control unit. During missions, the commander normally observes the battlefield with the compact sensor head.

This view shows the hatches of the commander and the scout to advantage. If it is not being used, the telescoping mast-mounted sensor is stored in its transport position to protect the sensitive sights.

The reliable belt-fed, air-cooled 7.62 x 51mm MG3A1 machine gun has an effective range of 1,200 meters and a muzzle velocity of 820m/s. It has a quick-change barrel that can be replaced after every 150 rounds fired.

Because of the limited space inside the narrow crew compartment, some personal equipment like the three sleeping bags, water, and numerous MREs (Meals, Ready to Eat) have to be stored in the large metal box attached to the front of the vehicle between the front lights.

The sprocket in detail. In the beginning, the Wiesel 1 Waffenträger versions were equipped with continuous rubber tracks, but these were later replaced by new Type 622 C steel tracks with replaceable rubber pads. These tracks are produced by Diehl Remscheid GmbH & Co. KG, the same as those used on the Aufklärungsfahrzeug Wiesel 1.

The stations of the commander and the scout are parallel to each other right behind the driver. To reduce weight and to provide an ergonomically comfortable seating position, specially designed seats from the German company of Autoflug GmbH were installed.

As with the Waffenträger Wiesel 1, the driver's station is located in the right front area of the hull. The hull itself consists of steel plates of varying thickness and consistency, providing the three-man crew sufficient protection against shrapnel produced by different weapons such as mines and artillery grenades, as well as from infantry ammunition.

Each of the crew seats, made of a very hardwearing green canvas material, is attached to the vehicle's roof with four lashing straps so that they have no direct contact with the floor plates. To provide the crew with an ergonomically comfortable position while observing and scanning the surrounding area through the open hatches, the mine protection seats can be locked in different positions by a sophisticated but very easy to use adjustment system attached to the roof. By pushing a switch, the soldiers can drop back into the safety of the crew compartment within seconds.

The Führungs- Kommunikations- und Informationssystem (FaKom - Command, Control and Communication System), built by Rheinmetall Defence Electronics, is located in the small space between the commander's and scout's stations. This very advanced C^3 system consists of a touch-screen display and control unit, a computer module behind the commander's seat, the data terminal MIDAT with a memory card, and a keyboard. For orientation in cross-country terrain, digitized maps are used that show the vehicle's position, target position, and the current battlefield environment.

Information is easily displayed as an icon on the FaKoM touch screen and the situation is continuously updated between vehicles and command post via a radio data link with HF and VHF radio. The AOZ 2000 control unit located below the FaKoM includes controls for brightness, contrast, and focus, as well as the magnification and focus switch for the thermal imager and the CCD day vision camera.

Aufklärungsfahrzeug Wiesel 1	
Combat Weight	3.23 tons
Payload	2.87 tons
Length	3.55m
Width	1.82m
Height to top of machine gun	1.84m
Height to top of sensor head	2.20m
Ground Clearance	30cm
Fording Depth	50cm
Trench Crossing	1.20m
Gradients	60%
Side Slopes	30%
Crew	3

Luftlandeaufklärungskompanie 310 Patch

A shield was chosen as the symbol for the internal unit patch, as this represents better than any other motif, the idea of being well fortified and stable. The larger upper part of the shield is gold in color while the lower part is light blue. The upper part symbolizes the connection with the Panzeraufklärer (Reconnaissance Troops) and Panzeraufklärungsbataillon 3 (Reconnaissance Battalion 3), to which the company belonged in peacetime up till 31 December 2005. The light blue in the lower part is the color of Luftlandebrigade 31 (Airborne Brigade 31), to which the company belongs during operations and with the new tactical organization that began on 1 January 2006. There are three symbols on the Luftlandebrigade 31 patch - a parachute and two crosses. The large motif on the gold background is explained as follows:

1. The open circle is part of the patch of Panzeraufklärungsbataillons 3 "Springender Reiter" (Springing Cavalier).
2. The designation of Luftlandeaufklärungskompanie 310 is integrated into the circle.
3. The eagle is the symbol of the German Bundeswehr Luftlandetruppen (Airborne Troops).

This new unit patch reflects bonds with the Panzeraufklärungstruppe, as well as the Luftlandetruppe. It is a symbiosis of elegance, power, firmness and self-assertion. Virtues that characterize both these branches are integrated, blended and concentrated in the new Luftlandeaufklärungskompanie 310.

The compact black and white sight display system mounted below the roof on the right side of the commander's seat, which can also be swiveled towards the scout if necessary, provides the observer with the ability to easily evaluate the reconnaissance images of both the thermal imager and the CCD day vision camera. Using a small, black joystick of the sensor head control unit located to the right of the commander's station, the sensor head can be easily controlled 360° in azimuth and -8° to +60° in elevation.

This photo shows the commander working with the FaKoM keyboard. The reliable radio system consists of a SEM 80/90 (VHF), a HRM 7400 (HF) and the HRS 7000 Manpack (HF) and SEM 52SL (VHF) handheld systems. Note the covered periscopes and the fire extinguisher in the background.

The TOW launcher consists of the traversing unit, the launch tube, the optical sight, the missile guidance system, the optical sight, and the AN/TAS 4 thermal imaging sight. Note that the mount of the reference sensor is part of the AGDUS (Ausbildungsgerät Duellsimulator) laser engagement system used at the Gefechtsübungszentrum Heer in Letzlingen, Germany.

This Waffenträger TOW Wiesel 1 provides fire support while paratroopers of Fallschirmjägerbataillon 263 enter the Hillerleben MOUT site in January 2006. The TOW can destroy all known armor at any distance between 65 and 3,750 meters. At the rear of the Wiesel 1 TOW's roof is one large hatch that can be folded to the rear to allow access to the six spare TOW missiles stored horizontally in the rear of the hull.

1/35 G6 155mm SPH

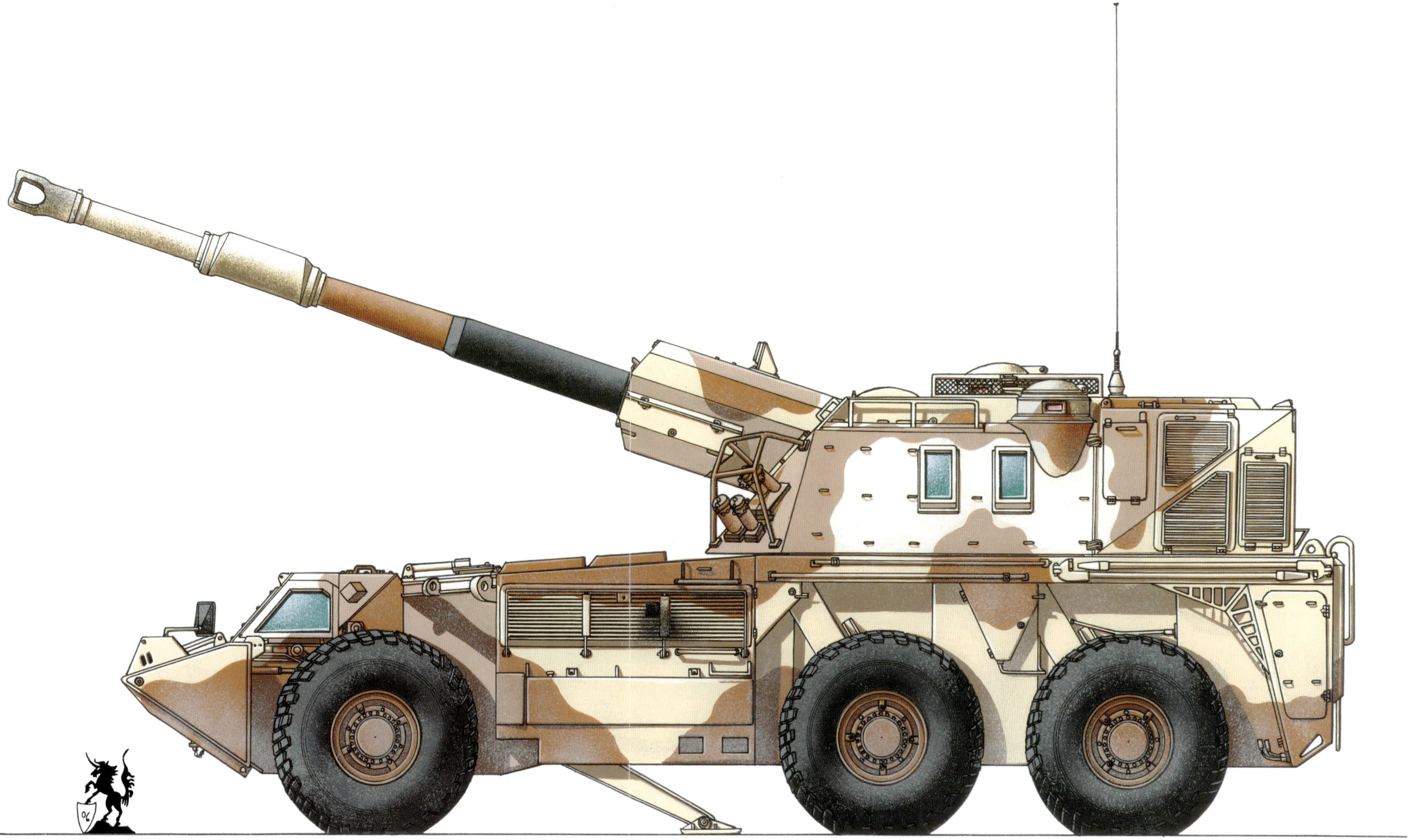

Hubert Cance

"Uhlan Eagle 2005"

1st (UK) Armoured Division trains in Poland

Carl Schulze

The OPFOR (Opposing Force) during Exercise "Uhlan Warrior 2005" consisted of an armored infantry company with Warriors and a tank squadron with Challenger 2 Main Battle Tanks (MBTs). The ten MTBs and their crews were supplied by B "Badger" Squadron of the 2nd Royal Tank Regiment, a unit that is based at Fallingbostel. In order to provide the OPFOR tanks with a slightly different appearance than their counterparts, they were fitted with empty Milan tubes instead of machine guns and a pyramid-type structure was mounted on their TOGS. The two white rings on the bore evacuator of the 120mm main gun of this tank identifies it as belonging to the 2nd Troop of the Squadron. The 1st Troop has one white ring, while the 3rd Troop has three white rings.

Explosions mark the incoming rounds of artillery fire. As clouds of white smoke drift through the air, the OPFOR (Opposing Force) soldiers of the 1st Battalion The Scots Guards dive for cover in their trenches. While the noise of the explosions is still echoing through the air, another noise can be heard - the sound of cannon and machine gun fire. The screaming noise of engines powering heavy armored vehicles and running at high speed can be heard; the noise gets louder and louder as the vehicles close in. Suddenly, the first FV510 Warrior IFVs (Infantry Fighting Vehicles) and Challenger 2 MBTs (Main Battle Tanks) appear out of a thick smokescreen. At high speed the armored fighting vehicles close in on the OPFOR positions. The Warrior turrets traverse left and right, the vehicles' chain guns firing without a break. An OPFOR soldier dressed in a desert DPM uniform appears from a trench, swinging an 84mm LAW80 to his shoulder. He takes aim at a Warrior that is just 80m away. The white smoke that erupts from the weapon indicates the launch of its deadly warhead. A split second later the TES (Tactical Engagement Simulation) equipment mounted on the infantry fighting vehicle begins to flash, an indication that the vehicle has been destroyed. The Warrior grinds to a halt and the crew turns the turret to the rear with the gun pointing up.

In the meantime, the Challenger 2 MBTs pass the positions of the OPFOR infantry at high speed, thrusting forward in order to establish a defensive screen well ahead of the fighting. It is the mission of the main battle tanks to prevent any enemy counter-attack that might be intended to stop the infantrymen and their Warriors from clearing the OPFOR positions. An FV510 Warrior passes an enemy trench by mere centimeters and stops five meters behind it. The rear door of the vehicle swings open and soldiers of the 1st Battalion The Princess of Wales's Royal Regiment (PWRR)

An FV510 Warrior IFV (Infantry Fighting Vehicle) passes over pipe fascines that were dropped by a Chieftain AVRE into an anti-tank ditch. Obstacle crossing was one of several complex scenarios that the Battle Groups had to master during Exercise "Uhlan Warrior". This Warrior belongs to the PWRR (The Princess of Wales's Royal Regiment) Battle Group. The FV510 Warrior has a crew of three that consists of the driver, gunner and vehicle commander. Up to seven fully equipped infantrymen can be transported in the vehicle's rear fighting compartment.

The Shielder Vehicle Launcher Scattered Mine System (VLSMS) is based on the flatbed version of the Stormer APC. The VLSMS is itself based on the U.S. Army's M163 Vulcan system, and it consists of 120 barrels, each of which contains six L34A1 anti-tank mines. A low-density minefield 360m wide and 2.7km long can be laid by the two-man crew at a speed of 75km/h. Here a Shielder VLSMS of 35 Engineer Regiment can be seen laying a minefield in support of the PWRR Battle Group.

dismount at high speed. The OPFOR men provided by the 1st Battalion The Scots Guards had stayed hidden inside their trench while waiting for precisely this moment - they raise their weapons and open fire. Orders are immediately barked out by the section commander of the dismounted troops from the Warriors, and instantly the three OPROR soldiers come under heavy fire. The PWRR troops throw smoke grenades while a fire team closes in on the trench. After one of the grenades thrown into the trench has detonated, one soldier of the team jumps into the position while firing his weapon. The OPFOR troops are all dead, so the soldier shouts to his comrades: "Trench clear! Three enemy dead!" But the fighting has not yet finished, as dismounted troops are busy clearing enemy trenches all over the area. This attack took place during "Uhlan Eagle 2005", an exercise conducted by the 1st (UK) Armoured Division at the Drawsko Pomorski Training Area (DPTA) in the autumn of 2005.

A Milan ATGW (Anti-Tank Guided Weapon) team of Anti-Tank Platoon, Y Fire Support Company, 1PWRR, defends its position during the "Hasty Defense" part of the Battle Group training. The Milan ATGW is fitted with the Tactical Engagement Simulation (TES) kit. The soldiers are amongst some 9,000 troops of the British Army that are from overseas countries. This particular team contains Fijians, of which there are 1,820 in British Army service. The Milan ATGW system has a combat range of 2,000m and it can penetrate up to 352mm of conventional steel armor.

"Uhlan Eagle 2005" in numbers

From 19 September to 25 November 2005, the 1st (UK) Armoured Division conducted Exercise "Uhlan Eagle 2005" at the Drawsko Pomorski Training Area (DPTA) in Poland. This exercise of the "First Division" was the largest deployment of troops from the Germany-based division since Operation Telic in 2003. Some 4,500 British soldiers with 956 wheeled and tracked vehicles deployed to Poland. The exercise also involved some troops that deployed from the UK to Poland, as well as Jaguar fighter aircraft that were temporarily based at a Polish Air Force base.

In addition, 250 Danish soldiers with 120 vehicles from the HQ of the 1st Danish Brigade deployed as part of the exercise to Poland. The British and Danish troop contingents deployed by road and rail to the exercise area. Altogether, the deployment and redeployment of troops, vehicles and stores involved thirty-two trains and seventy road convoys. In the case of the participating British troops, the convoys traveled some 750km to the DPTA from their home bases in Germany. The British Army has conducted large-scale exercises on the DPTA in Poland since 1996. The benefit of the DPTA is that, due to the size of the training area, Battle Groups up to "2/2" size can be exercised. A 2/2 Battle Group features two tank squadrons and two armored infantry companies, as well as all associated combat support assets. In addition, the DPTA features a lot of natural obstacles such as water obstacles and dense woods, which make the execution of tasks more challenging. "Uhlan Eagle 2005" was the first large-scale exercise conducted in Poland by the British Army since 2000. In actual fact, "Uhlan Eagle 2005" was not a single exercise but more of a series of exercises involving highly demanding training challenges for the operationally experienced troops of 20 Armoured Brigade and other participating units. To be more specific, these exercises were "Uhlan Warrior 2005", "Uhlan Barbara 2005", "Uhlan Kestral 2005" and "Rhino Charge 2005".

Portrait of a soldier from the 1st Battalion The Scots Guards. The unit's C Company provided the armored infantry element of the OPFOR. The unit belongs to 4 Armoured Brigade. Note the brigade's badge, featuring a black desert rat, worn by this soldier on his left upper sleeve.

Exercise "Uhlan Warrior 2005"

Exercise "Uhlan Warrior 2005" took place between 28 September and 6 November. The exercise consisted of three Battle Group exercises that each lasted twelve days. In every twelve-day period, one of the three Battle Groups of 20 Armoured Brigade from Paderborn, Germany, was trained to operate as a combined arms team in a full-scale war-fighting scenario. The BG exercises were the highlight of the training year for the brigade, as BGs were put through various operational scenarios involving defensive and offensive tasks such as Advance to Contact, Hasty Defense, and Fighting Withdrawal. The BGs participating in these exercises were the Queen's Royal Hussars BG (2/2), the 1st Battalion The Light Infantry BG (1/2), and the 1st Battalion The Princess of Wales's Royal Regiment BG (1/2).

With its Chieftain AVLB, the bridging section of 37 Armoured Squadron, 35 Engineer Regiment, is supported by a Tank Bridge Transporter (TBT). In this case, the TBT carries two additional 13.5m long No. 12 bridges. The TBT is based on the Alvis/Unipower M Series 8x8 truck. The vehicle is powered by a Cummins six-cylinder engine that allows it to reach a top speed of 84km/h. It is able to pass over terrain that is not usually suitable for wheeled vehicles.

Though three Battle Groups were trained during the exercise period, only one set of vehicles was deployed to Poland. The vehicles were handed over from one Battle Group to another, and altogether the Battle Group and the OPFOR fielded a total of twenty-five Challenger 2 MBTs and fifty-nine Warriors of all variants, all of which were brought by train to Poland. The vehicle set included armored fighting vehicles from every battalion that was exercising in Poland. The vehicles were deployed prior to the exercise series and joined together to form the complete set of vehicles as required.

The BG exercises were divided into two parts. The first half of the BG exercises focused on integration and interaction between the different elements of the BG. In the second part, the BG was exercised by 20 Armoured Brigade HQ, and had to execute missions given by the brigade commander, Brigadier Nick Carter CBE. Exercise "Uhlan Warrior 2005" involved the following units:
1st Battalion The Light Infantry
1st Battalion The Princess of Wales's Royal Regiment
20 Armoured Brigade HQ and Signal Squadron
The Queen's Royal Hussars
35 Engineer Regiment
3rd Battalion REME
1st Battalion REME
28 Engineer Regiment
58 Battery, 12 Air Defence Regiment Royal Artillery
TAC Group, 26 Regiment Royal Artillery
1st Battalion The Scots Guards
2nd Royal Tank Regiment

Exercise "Uhlan Eagle 2005" also saw a couple of new Oshkosh/ABRO Close Support Tankers being deployed. The tractor-truck is based on the Oshkosh MTVR truck, which is widely used by the United States Marine Corps. The semi-trailer of the Close Support Tanker can hold 20,000 liters of fuel. Powering the truck is a Caterpillar C-12 engine that develops 425hp. The pictured vehicle belongs to a unit of 102 Logistics Brigade.

A sniper team of the 1st Battalion The Princess of Wales's Royal Regiment is in position. Note that the 7.62mm L96A1 sniper rifle is fitted with TES equipment. TES allows realistic training for combat operations by using laser technology. Weapon-mounted devices emit a signal that is collected by sensors mounted on vehicles and personnel, and noise, smoke or flashing lights indicate to the crew or wearer that a hit has been registered.

This Stormer HVM belongs to 58 Battery, 12 Air Defence Regiment Royal Artillery. The Stormer HVM is armed with the Starstreak High Velocity Missile system, which has a combat range of 5.5km. Up to eight missiles can be stored in a ready-to-launch position in the launcher. The Air Defence Alerting Device (ADAD) is situated on top of the launcher. Also visible is the Commander's Acquisition System (CAS) located in front of the commander's hatch.

1st Battalion The Princess of Wales's Royal Regiment Battle Group

"Assault Journal" visited Exercise "Uhlan Eagle 2005" during the last "Uhlan Warrior" Battle Group exercise, which featured the 1st Battalion The Princess of Wales's Royal Regiment Battle Group. In the following text, we will provide our readers with a guided tour around the BG. The major portion of troops in the BG came from the 1st Battalion The Princess of Wales's Royal Regiment. The PWRR currently consists of three battalions, with one of these, the 3rd Battalion, being a Territorial Army unit. With more then 550 Battle Honours, the regiment is the most decorated in the whole of the British Army. Soldiers of the regiment have won fifty-seven Victoria Crosses, the most recent one being won by Private Johnson Gideon Beharry in Iraq in 2004.

The nickname of the regiment is "The Tigers". It stems from the 67th Regiment of Foot (South Hampshire Regiment), one of the regiments of which the PWRR maintains a history. Having served twenty-one years of unbroken service in India under active service conditions, King George IV authorized the Indian Tiger sleeve badge for the 67th Regiment in 1826. Today the badge is still worn by all ranks in the regiment. The 1PWRR provided the bulk of the BG troops, and the 1PWRR HQ Company became the BG HQ. The company consisted of the A1 and A2 Echelons, the Regimental Aid Post, the HQ Light Aid Detachment, as well as the BG HQ Main A and BG HQ Main B. Y Company, the 1PWRR's Manoeuvre Support Company, supported the BG with its Reconnaissance Platoon of eight FV107 Scimitar CVR(T)s, its Assault Pioneer Platoon with three

Front view of a Challenger 2 MBT of The Queen's Royal Hussars. Parts of the TES kit can be seen in and under the 120mm gun barrel. A part of the kit is the loudspeaker that is visible to the right of the gun on the turret's front armor, as well as the flare system on the left side of the gun and the sensor box in the center of the turret.

A Challenger 2 MBT of The Queen's Royal Hussars in an over-watch position. The tank has a combat weight of 64 tons. Its crew consists of the driver, the vehicle commander, the gunner and the loader - four men in total. The vehicle is 11.753m long when the gun is in the 12 o'clock position, and it is 3.038m high. The vehicle is able to cross trenches up to 2.8m wide and it can ford water up to 1.07m deep.

FV510 Warrior IFVs, its Mortar Platoon with nine 81mm L16A2 mortars mounted in FV432APCs, and its Anti-Tank Guided Weapon Platoon with eight FV510 Warrior IFVs and sixteen Milan ATGWs. Of course, the 1PWRR also provided the armored infantry companies of the BG, namely A and B Companies. Each of the two companies fielded twelve FV510 Warrior IFVs and two FV511 Warrior Infantry Command Vehicles. The companies were organized into the Company HQ element and three infantry platoons with four Warriors each.

A platoon consisted of some thirty-five soldiers divided into three ten-man sections and the platoon HQ. The infantrymen were armed with the 5.56mm SA80A2 assault rifle, the 5.56mm LSWA2 light machine gun, the 7.62mm GPMG machine gun, and the 84mm LAW80 anti-tank weapon. Heavy firepower for the BG was provided by the 120mm guns of Challenger 2 MBTs from B Squadron of The Queen's Royal Hussars. The squadron fielded fourteen Challenger 2s, which were divided among the squadron's four troops, each with three tanks, and the Squadron HQ with two tanks. Engineer support for the BG was provided by 37 Armoured Squadron and parts of 44 HQ Squadron of 35 Engineer Regiment. The engineers fielded two Chieftain AVLBs, two Chieftain AVREs, two Shielder VLSMS vehicles, one Tank Bridge Transporter, an Automotive

During the "Uhlan Warrior" exercises, the Battle Groups and their activities were closely monitored by the Brigade Commander of 20 Armoured Brigade. The Brigade Commander traveled around the battlefield in an FV511 Warrior Infantry Command Vehicle that could easily be recognized by the large sticker featuring the badge of 20 Armoured Brigade. The soldier in the picture is not the Brigade Commander, but is the vehicle commander, who always had to make sure that the Warrior was at the required spot at the correct time.

Of the three Battle Groups taking part in Exercise "Uhlan Warrior", two were 2/1 BGs, which means that they consisted of two armored infantry companies and one tank squadron. Namely these BGs were the 1LI and 1PWRR BGs. The third BG was the QRH BG, which features two tank squadrons and two armored infantry companies. Here an FV510 IFV used by the PWRR BG can be seen. Note the camouflage made of sandbags that have been cut into strips.

Posing for the camera, this soldier of the 1st Battalion The Princess of Wales's Royal Regiment shows the unit's cap badge and flash. With more then 550 Battle Honours, the regiment is the most decorated in the whole British Army; its nickname is "The Tigers". Soldiers of the regiment have won fifty-seven Victoria Crosses, the most recent one being awarded to Private Johnson Gideon Beharry in Iraq in 2004.

Bridge Launching Equipment (ABLE) system, two Combat Engineer Tractors, and several FV432 and FV103 Spartan armored personnel carriers, amongst other assorted equipment.

16 Battery of 26 Regiment Royal Artillery supported the BG with artillery support. While the gunfire was simulated and no fire units were deployed, the Artillery Tactical Group was still deployed in the field. This consisted of the Battery Commander and his staff providing the BG's Offensive Support Cell. Also deployed were three Forward Observation Officer (FOO) teams with FV514 Warrior Mechanised Artillery Observation Vehicles (MAOV). The artillery soldiers directed and coordinated the simulated indirect fire and close air support for the BG. Close air-defense for the BG was provided by an Air Defence Fire Group of 58 Battery, 12 Air Defence Regiment Royal Artillery. This unit fielded four Stormer HVM vehicles and a FV4333 Stormer Troop Reconnaissance Vehicle (TRV). Logistics support for the battle group was provided by various units, but since these were not an integral part of the BG, they are not covered in this article.

Exercise "Rhino Charge 2005"

Exercise "Rhino Charge 2005" occurred from 7-11 November 2005. The exercise was designed to train the HQ of 20 Armoured Brigade in working under the HQ of the 1st (UK) Armoured Division in a war-fighting context. The exercise was a Command Post Exercise (CPX) and saw only the HQs moving around the DPTA. During the exercise, the deployed Brigade HQs had to plan several complex defensive and offensive operations, of which some had to be executed simultaneously. Other troop movements and combat activities were simulated by the Advanced Battlefield Computer Simulation System (ABACUS). For the exercise, ABACUS (which is usually based at the Land Warfare Collective Training Group [LWCTG] in Sennelager Germany) was deployed forward to the DPTA and set up in a huge tent complex. Also deployed for the exercise were HQ 4 Armoured Brigade and the HQ of the 1st Danish Brigade. The Danish element taking part in the exercise involved 250 personnel and 120 vehicles. During the exercise, the Danish troops conducted trials for new setups for their Brigade HQ.

Soldiers of C Company, 1st Battalion The Scots Guards, dismount from their FV510 Warrior IFV while attacking a position of the 1PWRR Battle Group. The 1SG belongs to the 4th Armoured Brigade and is currently based in Münster, Germany. During Exercise "Uhlan Warrior 2005", this company provided part of the OPFOR. Note that the soldiers wear Desert DPM uniforms to make identification easier.

In preparation for a bridging mission, engineers of 35 Engineer Regiment conduct a dry run with their Automotive Bridge Launching Equipment (ABLE). In the background, the bridge can be seen already pushed over the simulated obstacle. In this case the bridge has a length of 32m. Each system is comprised of one ABLE vehicle (as seen on the right) and two Bridging Vehicles (BV) (of which one can be seen on the left).

After they have dismounted from their Warrior IFVs, infantrymen of the 1PWRR search a wooded area for enemy forces. The pictured soldier is armed with a 5.56mm SA80A2 assault rifle. Contact between the troops is maintained with their Personal Role Radios (PRR). Note the TES kit that is worn by the soldier and that is also mounted on his weapon.

Exercise "Uhlan Barbara 2005"

Exercise "Uhlan Barbara 2005" was the final part of the "Uhlan Eagle 2005" exercise series, and it took place from 12-20 November 2005. The exercise was a live-fire exercise of the 1st (UK) Armoured Division Offensive Support Group (OSG), an event that is conducted biannually. This year the main aim of the exercise was to develop integration procedures for surface and air delivered effects in a high-intensity conflict. In addition, the participating units trained in the integration of air-defense assets and close air support missions. The exercise included live firing of six 155mm AS90 self-propelled howitzers of 26 Regiment Royal Artillery, and the live firing of six MLRS rocket launchers of 39 Regiment Royal Artillery. In order to field more fire units that needed to be coordinated, the guns and rocket launchers deployed to the DPTA in pairs or as gun platoons of three weapon systems. Additionally, Jaguar fighter aircraft of the RAF dropped bombs on the DPTA. "Uhlan Barbara 2005" involved the following units:

4 Armoured Brigade HQ and Signal Squadron
16 Air Defence Regiment Royal Artillery with Rapier Field Standard B
12 Air Defence Regiment Royal Artillery with Stormer HVM
Phoenix TAC Group of 32 Regiment Royal Artillery
1 Regiment Army Air Corps with five Lynx Mk 9 LBH
618 Tactical Air Control Party
619 Tactical Air Control Party
TAC Groups of 4 Regiment Royal Artillery.
39 Regiment Royal Artillery with six MLRS
26 Regiment Royal Artillery with six AS90 self-propelled howitzers

Along with the exercise troops, instructor teams of the Sennelager-based Land Warfare Collective Training Group (LWCTG) deployed to Poland. The LWCTG supported the exercise with a pool of umpires, and it worked hard to make the training scenarios as realistic as possible. The umpire teams were omnipresent in their Land Rovers marked with bright red panels. Here a Land Rover 110XD FFR used by the umpires can be seen.

After a ditch has been bridged by pipe fascines and a berm has been bulldozed away, Challenger 2 MBTs advance through the breach made in the enemy's obstacle belt. These tanks belong to B Squadron, The Queen's Royal Hussars, based at Paderborn. This regiment is the only tank unit within 20 Armoured Brigade. The Challenger 2 tank is armed with a 120mm rifled L30A1 tank gun. As a coaxial weapon, a 7.62mm L94A1 Hughes Chain Gun is mounted next to the gun. For close defense and air-defense purposes, a 7.62mm GPMG is mounted on the turret roof near the loader's hatch.

The FV510 Warrior is the primary infantry fighting vehicle of the British Army. The vehicle has a combat weight of 25.4 tons, is 2.74m in height, 3.10m in width, and 6.58m in length. The Warrior is powered by an eight-cylinder turbocharged diesel engine. The vehicle can reach a top speed of 75km/h and has a road range of 600km. This FV510 Warrior is a section vehicle used by the 1PWRR BG during Exercise "Uhlan Eagle 2005".

A Chieftain AVRE bulldozes a gap through a berm in an enemy obstacle belt. The vehicle belongs to 37 Armoured Squadron, 35 Engineer Regiment. Instead of the Track Width Mine Plough, the vehicle is equipped with the Bulk Earth Moving Attachment (BEMA) needed to fulfill its task. Of the three Chieftain AVREs in the Armoured troop of the squadron, usually two are equipped with the mine plough and the third is fitted with the BEMA. However, both mounts are available for each vehicle and usually the one that is not mounted is carried on the vehicle's rear storage area.

This lance corporal belongs to the Armoured Troop of 37 Armoured Squadron, 35 Engineer Regiment. This soldier is the vehicle commander of a Chieftain AVRE. Note the Royal Engineers flash and shoulder title worn on his right upper sleeve.

Exercise "Uhlan Kestral 2005"

Exercise "Uhlan Kestral 2005" took place from 19 September to 25 November 2005. Basically, "Uhlan Kestral 2005" encompassed all logistics efforts that were necessary to make the "Uhlan Eagle 2005" series of exercises happen successfully. This included all efforts to deploy and re-deploy British troops to Poland and to sustain them. The deployment to Poland of 20 Armoured Brigade and 1st (UK) Armoured Division elements provided a good opportunity to train the logistics units for expeditionary operations. A Movement Control Centre (MCC) monitored and coordinated all convoy movements. Convoys were assembled at a Convoy Mounting Centre that was established near Hannover in northern Germany. From here the convoys moved 700km east to the DPTA. The journey involved several stops for refueling and maintenance, as well as the crossing of the Germany-Poland border. Also, part of the exercise was the in-time delivery of food, fuel and ammunition stores to exercise troops. Participating in the exercise were units of 102 Logistics Brigade, based in Germany, plus units of the Division Support Group.

Busy times

For the highly trained soldiers of the 1st (UK) Armoured Division, the autumn of 2005 was a busy time. While 20 Armoured Brigade, supported by units of 4 Armoured Brigade and other elements of the division, were proving their combat readiness in Poland, other exercises were being run elsewhere by the remaining troops of the division. The 1st (UK)

As the sun disappears over the horizon, an engineer keeps watch while his Battle Group crosses an obstacle. The soldier is manning a 7.62mm GPMG machine gun mounted on the commander's hatch ring mount of an FV432 APC.

For heavy recovery work, the Light Aid Detachments (LAD) of armored regiments are equipped with the Challenger Armoured Repair and Recovery Vehicle (CRARRV). The vehicle has a crew of three, which consists of a commander, driver and radio operator. Main recovery equipment of the vehicle includes a front-mounted hydraulically operated dozer blade, a hydraulically driven double capstan winch with 150m of rope, and a 6.5-ton crane.

Armoured Division Offensive Support Group, with most of its artillery assets, conducted the live-fire exercise "Shelldrake Sword 2005" on the Munster and Bergen Hohne Ranges. The exercise also involved MLRS vehicles of 39 Regiment Royal Artillery from the UK. For a period of four weeks, guns and rocket artillery systems conducted extensive live-fire training. In southern Germany, the 1st Queen's Dragoon Guards conducted the four-week long reconnaissance exercise "Swabian Eagle 2005", together with German reconnaissance units. On top of this, the operationally experienced 7 Armoured Brigade "Desert Rats", as well as parts of the 1st (UK) Armoured Division HQ, deployed to Iraq for Operation Telic VII in October 2005.

Warrior FV510 IFVs of the 1PWRR advance to contact during an early stage of the Battle Group Exercise "Uhlan Warrior". The Warrior's main armament is a 30mm RARDEN cannon. As its coaxial weapon, an L94A1 Chain Gun is mounted to the left of the RARDEN cannon. An armored infantry company can field twelve Warrior FV510 IFVs and two Warrior FV511 Infantry Command Vehicles.

In order to allow the exercise troops to cross the several water obstacles found in the DPTA, a troop from 23 Amphibious Squadron, 28 Amphibious Regiment, based at Hameln, deployed with their eight M3 amphibious rigs to Poland. Here a ferry formed from four M3s can be seen. Such a ferry can transport three Warrior IFVs or two Challenger 2 MBTs. In this case the load consists of a Foden recovery vehicle and a Warrior IFV.

An FV103 Spartan passes through a water obstacle in the Drawsko Pomorski Training Area. The vehicle belongs to 37 Armoured Squadron, 35 Engineer Regiment. The engineers use the Spartan as a troop carrier, reconnaissance vehicle, and transport vehicle for troop and squadron command elements. The vehicle is powered by a Cummins diesel engine and has a combat weight of 8.7 tons.

The hard training in Poland took its toll on troops and equipment. In the first two Battle Group exercises, a total of forty-eight power packs had to be changed. Here an FV512 Warrior Mechanised Combat Repair Vehicle (MCRV) can be seen extracting the engine of an FV510 Warrior Infantry Section Vehicle. The FV512 Warrior MCRV is equipped with a hydraulic crane capable of lifting 6.5 tons. Inside the vehicle, a fully equipped workshop, including power tools and a welding kit, is carried.

20 Armoured Brigade – A short history

Today's 20 Armoured Brigade was originally formed as 20 Light Armoured Brigade in East Anglia on 2 September 1939. At that time it was a Territorial Army unit and the brigade's insignia was a white armored head of a horse. The first units belonging to the brigade were Yeomanry Regiments that included the Royal Gloucestershire Hussars and the Northamptonshire Yeomanry. 20 Light Armoured Brigade was renamed 20 Armoured Brigade on 14 April 1940, the name it has retained up to this day. The brigade became part of the 6th Armoured Division when this tank formation was established on 12 September 1940. By then the insignia of the 6th Armoured Division had changed to a white "Armored Gauntlet" on a black background. The mailed fist was selected by the Division as a symbol of the hard punch that an armored formation delivers against the enemy. On 30 April 1943, 20 Armoured Brigade was disbanded. The 6th Armoured Division deployed to North Africa where it fought at Kasserine, Fondouk, El Kourzia and Tunis in 1943. The years of 1944 and 1945 saw the division in action in Italy fighting at Monte Cassino, the Gothic Line and the Argenta Gap. When combat ended in Europe on 8 May 1945, the 6th Armoured Division was in Austria.

The post-war years saw a reshuffle of the British Army, which included the reforming of 20 Armoured Brigade on 15 September 1950. In December 1951 the brigade moved to Münster in Germany, and from 1952 onwards, it again formed part of the 6th Armoured Division. In April 1958 the 6th Armoured Division was disbanded and the surviving 20 Armoured Brigade came under command of the 4th Armoured Division. For traditional reasons the brigade assumed the fist badge of the division, although the background for the white fist changed from black to blue. From 1974 to 1977, the brigade operated under the contemporary task force concept. At this time it was called Task Force Hotel, but it regained its original name soon after.

With the end of the Cold War and the cutting of the presence of British forces in Germany in the early nineties under the "Options of Change" policy, the brigade moved to its current home base at Paderborn and came under control of the 1st (UK) Armoured Division. In April 1995, 20 Armoured Brigade deployed to Bosnia as part of UNPROFOR. During its deployment, the brigade was involved in a lot of incidents that ultimately led to the peace declaration and the beginning of the NATO-led IFOR mission. From October 1996 to April 1997, the brigade deployed for a second time to Bosnia. Based at Sipovo and later at Banja Luka, the brigade transitioned from IFOR to SFOR. Under the ambit of SFOR, the brigade again deployed to Bosnia between August and December of 1999.

Between November 2003 and April 2004, 20 Armoured Brigade deployed for the first time to Iraq as part of Operation Telic. During this period, the brigade commanded the Danish Contingent (DANCON) (which also included a Lithuanian infantry platoon) in addition to its own units. In the spring of 2005 the brigade returned to Iraq to again serve in the Multi-National Division (South-East).

With its crane, a Bridging Vehicle (BV) lifts part of a General Support bridge on the Automotive Bridge Launching Equipment (ABLE). During Exercise "Uhlan Eagle 2005", the Combat Support Troop of 44 HQ Squadron, 35 Engineer Regiment, supported the exercising Battle Groups, as well as the OPFOR, with its ABLE system in order to provide vital obstacle crossing capabilities.

A Challenger 2 MBT moves to a new position. The vehicle belongs to B Squadron, The Queen's Royal Hussars. At the time of writing, the regiment could field four tank squadrons, with each of these squadrons possessing fourteen Challenger 2 tanks. Future plans see the number of tanks per squadron being raised, while one squadron of the regiment will be converted into a Medium Armour Squadron. The Challenger 2 is powered by a Perkins Condor V12 twin-turbocharged diesel engine that develops 1,200bhp, which allows the vehicle to reach an average cross-country speed of 40km/h.

Soldiers of C Company, 1st Battalion The Scots Guards, have just mounted up in their FV510 Warrior IFV. In the rear fighting compartment of the Warrior is space for seven fully equipped infantrymen. The troops sit on padded two-man benches, with two benches situated on the left side and a third on the right. The seventh seat is also on the left; it can also be lifted up since a chemical toilet is situated underneath it.

Each armored infantry regiment and armored regiment of the British Army can field a reconnaissance platoon. The platoon is equipped with eight FV107 Scimitar Combat Vehicle Reconnaissance (Tracked). The Scimitar CVR(T) is armed with a 30mm RARDEN cannon and a coaxial 7.62mm machine gun. The Scimitar has a crew of three that consists of the driver, gunner and vehicle commander. The vehicle has a combat weight of 7.8 tons, and is powered by a Cummins diesel engine that propels it to a top speed of 80km/h.

This FV4333 Stormer Troop Reconnaissance Vehicle (TRV) belongs to the Battle Group's air-defense fire group leader. "Uhlan Eagle 2005" saw air-defense fire groups of 58 Battery, 12 Air Defence Regiment Royal Artillery, providing air-defense cover for the exercising Battle Groups. The Stormer TRV has a combat weight of 12.7 tons and is powered by a Perkins six-cylinder engine that develops 250bhp, and propels the vehicle to a top speed of 80km/h.

Here a Challenger 2 MBT of B Squadron, The Queen's Royal Hussars, can be seen during a PWRR Battle Group attack on an enemy position. With the equipment being handed over from battle group to battle group, the number of heavy vehicles deployed to Poland was kept as few as possible. Only twenty-five Challenger 2 tanks were deployed for the exercise. Therefore, it should be noted that most of the vehicles used by the exercise troops did not belong to their unit's organic vehicle fleet.

The British Army uses the GKN Defence T 4 High Mobility Trailer in conjunction with the FV512 Warrior MCRV. One of them can be seen here hitched behind an MCRV. The trailer is equipped with quick-release mounting points and it can carry one Challenger or two Warrior power packs. It is fitted with an air brake system, and fully loaded, the trailer has a weight of 10.5 tons. Unloaded, it has a total length of 4.94m, a width of 2.89m, and a height of 1.97m.

Also deployed on Exercise "Uhlan Eagle 2005" was the Automotive Bridge Launching Equipment (ABLE). The ABLE systems belong to the Combat Support Troop of 44 HQ Squadron, 35 Engineer Regiment. The troop consists of two ABLE systems. Each system comprises one ABLE vehicle (center) that basically launches the bridge, and two Bridging Vehicles (BV) (left and right) that carry the bridge parts. A basic AVLB system with its bridge parts can establish a 32m long General Support Bridge that can take tracked loads up to MLC 70 and wheeled loads up to MLC 110.

An FV432 Mortar Carrier of Mortar Platoon, Y Company, 1PWRR, moves to a new firing position. An 81mm L16A2 mortar is mounted in the rear fighting compartment of the FV432. The weapon has a combat range of over 5km and it can fire High Explosive, Illumination or White Phosphorus Smoke grenades. The FV432 Mortar Carrier has a combat weight of some 15 tons, and it can reach a top road speed of 52km/h.

A Challenger 2 MBT of B Squadron, The Queen's Royal Hussars, advances through a ford in the Drawsko Pomorski Training Area. The tank has a ground clearance of 512mm, plus it can cross vertical obstacles up to a height of 900mm.

A Chieftain Armoured Vehicle Royal Engineers (AVRE) breaches an enemy minefield by using its front-mounted Pearson Track Width Mine Plough. The vehicle belongs to the Armoured Troop of 37 Armoured Squadron, 35 Engineer Regiment. In total, the troop can field three Chieftain AVREs and three Chieftain AVLB bridge-layers. Soon, such Chieftain-based engineer vehicles will be replaced by Trojan and Titan, vehicles that are based on the Challenger 2 chassis.

A rear view of a Chieftain Armoured Vehicle Royal Engineers of 37 Armoured Squadron, 35 Engineer Regiment. Note the Track Width Mine Plough stored on the rear deck of the vehicle. A bundle of pipe fascines is stored on the front deck.

From the commander's cupola of an FV103 Spartan APC, a member of an engineer reconnaissance party observes an enemy obstacle. He is using a pair of Leica Vector binoculars with integrated laser rangefinder. The soldier belongs to 37 Armoured Squadron, 35 Engineer Regiment.

During a fighting withdrawal, engineers of 37 Armoured Squadron, 35 Engineer Regiment, place explosives on an AVLB tank bridge. The bridge will be blown after the last vehicle of the 1PWRR Battle Group has passed over it. With the bridge blown, the enemy has no way of crossing the obstacle belt set up by the Battle Group. In this way, its attack is stopped even before it can gain momentum.

20 Armoured Brigade and its three Battle Groups, based in Paderborn, Germany, were the main focus of Exercise "Uhlan Eagle 2005". The badge of the brigade is an iron gauntlet, which mirrors well the brigade's capacity to dominate the battlefield by conducting armored thrusts into an opposing enemy force.

The FV432 has been in service as an armored personnel carrier with the British Army since the early 1960s. In infantry units it has been replaced by the Warrior IFV, but other branches of the British Army still use the elderly vehicle. This FV432 belongs to 37 Armoured Squadron, 35 Engineer Regiment, where it functions as an engineer section vehicle. The FV432 has a combat weight of 15.3 tons, and it is powered by a K60 six-cylinder diesel engine that produces 240bhp. It can reach a maximum speed of 52km/h.

During Exercise "Uhlan Eagle 2005", 652 Squadron "Frogs", 1st Regiment Army Air Corps (based in Gütersloh, Germany), was tasked with providing helicopters for CASEVAC missions in case of accidents. In addition, the five Lynx Mk 9 LBHs of the squadron provided an air transport capability for the Battle Groups during the "Uhlan Warrior" exercises, as well as acting as Air Observation Posts during Exercise "Uhlan Barbara".

Standing at the helmsman's controls, an engineer of 23 Amphibious Squadron, 28 Engineer Regiment, guides an M3 ferry across a lake in Poland. A single troop of 23 Amphibious Squadron can field eight M3 rigs, which can either be used as ferries or employed to build a bridge up to 100m long. A ferry with an MLC 70 payload can reach a top speed of 11km/h. Note the engineer flash worn by the soldier on the upper right sleeve of his Combat 95 pattern DPM jacket.

An OPFOR Challenger 2 MBT of the 2nd Royal Tank Regiment blocks a track in the Drawsko Pomorski Training Area. Clearly visible are the modifications that were instituted to make the vehicle easier to distinguish from tanks belonging to the exercise troops. Note the open door of the Thermal Observation and Gunnery Sight II (TOGS II), which is situated above the 120mm main gun.

A column of Warriors advances through a breach in an enemy minefield. The leading vehicle is an FV511 Warrior Infantry Command Vehicle. The vehicle has a crew of six, consisting of driver, commander, gunner and three radio operators. The only distinguishing feature of the FV511 from the FV510 when viewed from the front is that the FV511 has four antenna mounts on the turret, whereas the FV510 has only two.

In order to make the FV107 Scimitar CVR(T) of the OPFOR look different during Exercise "Uhlan Warrior 2005", long pipes were placed over the 30mm RARDEN cannon.

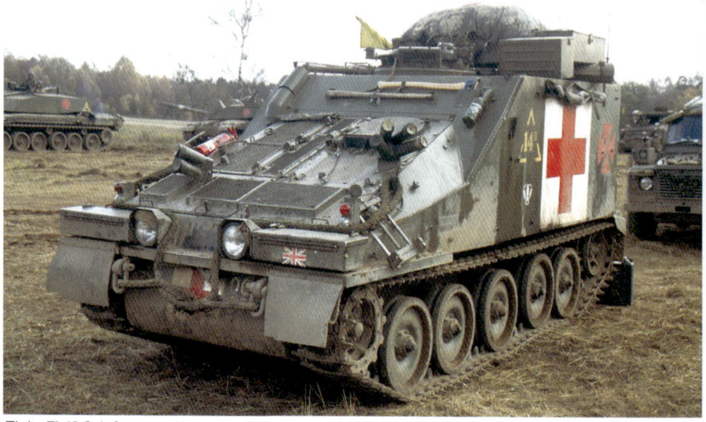

This FV104 Samaritan belongs to B "Badger" Squadron, 2nd Royal Tank Regiment. The Squadron provided the tanks used by the OPFOR Battle Group during Exercise "Uhlan Warrior". The Samaritan is based on the CVR(T) chassis and it can transport up to six wounded. Note the badger painted on the vehicle's side. A similar badge is also painted on the squadron's Challenger 2 MBTs.

An OPFOR FV510 Warrior IFV during an attack on positions of the PWRR Battle Group. The armored infantry component of the OPFOR was provided by C Company, 1st Battalion The Scots Guards. This company deployed ten Warriors to Poland. Note the Milan tube that is mounted on the turret, a practice used to make the vehicles look different to Warriors of the PWRR Battle Group.

Two Stormer HVMs of 58 Battery, 12 Air Defence Regiment Royal Artillery, move to a new firing position. During the "Uhlan Warrior" Battle Group exercises, the battery provided close air-defense for the ground forces. Adding an air-defense asset to a battle group makes the planning and execution of combat operations more complex. The Stormer HVM is armed with the Starstreak High Velocity Missile (HVM) system. The HVM missile has a combat range of 5.5km.

This Challenger 2 MBT belongs to the Commanding Officer (CO) of B Squadron, The Queen's Royal Hussars. The tank has a crew of four, including the commander, driver, gunner and loader. In evidence is the tank's running gear, which consists of a rear drive sprocket, a front idler, three return rollers and six large road wheels. The track is a double pin type with a width of 650mm; its tension can be hydraulically adjusted.

This Shielder Vehicle Launcher Scattered Mine System is one of three such vehicles that belong to 35 Engineer Regiment. In wartime, the unit would receive another Shielder VLSMS so that it can form two troops, each with two vehicles. The vehicle has a combat weight of 13.5 tons and is powered by a 250hp diesel engine that is Euro 2 compliant. The chassis of the Shielder is configured as a flatbed load carrier. The mine scattering system is mounted on the flatbed. The crew is located in the front part of the vehicle under full armor and NBC protection.

This soldier belongs to a troop of the 1st Battalion The Princess of Wales's Royal Regiment, which secured a ferry crossing site during a fighting withdrawal. He is armed with a 5.56mm SA80A2 assault rifle fitted with a TES kit. Note the "Tiger" flash worn on the upper right sleeve. The nickname "The Tigers" comes from the 67th Regiment of Foot (South Hampshire Regiment), with the PWRR maintaining the history of this regiment. Having served twenty-one years of unbroken service in India while under active service conditions, King George IV authorized the Tiger sleeve badge in 1826. Today the badge is still worn by all ranks in the regiment.

A Chieftain AVLB with a No. 12 bridge of 37 Armoured Squadron, 35 Engineer Regiment, closely follows the PWRR Battle Group during its advance. In a case where an obstacle is encountered, the vehicle will quickly deploy forward, lay its bridge and thus allow the BG to continue its advance. With the 13.5m long No. 12 bridge that is pictured here, gaps up to 12m wide can be bridged. When loaded with the No. 12 bridge, the Chieftain AVLB has a combat weight of 46.8 tons.

Belgian Armored Reconnaissance

Modernizing for the Future

Tim Mätzold, Daniel Nowak and Clemens Nienser

A Scimitar CVR(T) belonging to A Eskadron, 1 JP, secures a road during Exercise "Yellow Bastard '99." During this exercise, 4 ChCh played the enemy against 1 JP. This crew seems to be expecting enemy contact. (Daniel Nowak)

Background

The roots of Belgian armored reconnaissance go back to the early days of mounted troopers. Military men recognized early on that successful prosecution of a battle requires the acquisition of information about the enemy's condition and intentions. This fact was even defined by King Frederick the Great in 1743 for the *Preussische Husarenregimenter* (Prussian Hussars Regiment). These very regiments were tasked with reconnaissance missions, as well as with fulfilling their basic cavalry duties.

This Bombardier Iltis of the Staff and HQ Section, 4 ChCh, is fitted with red markings during a tactical exercise, identifying it as an OPFOR (Opposing Forces) vehicle. These vehicles are often used for playing the enemy to train all combat units, especially on low-level exercises. (Tim Mätzold)

When horses first began to be used for military purposes, it soon became obvious that apart from being good transport animals, they were also ideal for carrying out reconnaissance missions. Their main advantage was their mobility, which enabled soldiers to negotiate varied terrain in a short period of time to deliver timely information to their commanders about the enemy's status.

Up till 2004, there were three active units that belonged to the Belgian armored reconnaissance force:
1 *Jagers te Paard* (1 JP)
2 *Jagers te Paard* (2 JP)
4 *Chasseurs à Cheval* (4 ChCh)
The names of these units were given based on their history. The names *Jagers te Paard* and *Chasseurs à Cheval* both mean "hunters on horseback", alluding to the military tradition of mounted cavalry. The difference between them stems from the French and Flemish language forms, which are simply a reflection of Belgian geography and demography. The country is divided into French-speaking and Flemish-speaking regions (including a small German-speaking area). These factors are reflected in their unit insignias.

Insignia of 4 ChCh

4 ChCh's insignia consists of the following symbols:
- Two crossed sabers
- The number "4" beneath the crossed swords between the crown and the horn.

These items represent the typical kit of a soldier mounted on a horse. The unit's war cry, "*De Geules et d'Or*" (referring to their unit colors), is positioned underneath,

This particular manner of fitting the vehicles with stowage bins, boxes, jerry cans, and personal kit is unique to Belgian reconnaissance units. The needs of each crew determine how the vehicles are equipped. This 4 ChCh Scimitar observes a main road from a short distance away in a somewhat concealed position. (Tim Mätzold)

Enemy lines have been broken! This recce troop, which is mounted on two FV107 Scimitars, secures access to a bridge over the Aller River. (Tim Mätzold)

showing their perpetual spirit. The war cry's meaning can be traced to the colors red and yellow. *De Geules* (red) shows the spirit of the cavalry, and *D´Or* (yellow) illustrates the loyalty of serving king and country.

Insignia of 2 JP

The main difference between the insignia of 2 JP and that of 4 ChCh is that green is used instead of yellow. The name "*Reigersvliet*" is positioned underneath the horn in remembrance of the successful battle at the *Reigersvliet* on 6 March 1918.

Insignia of 1 JP

1 JP's insignia consists of the basic colors of green and yellow, two crossed swords, and a horn. The unit's war cry, "*Ubique Fidelis et Fortis*," is located underneath. The words have the following meaning:

Ubique: Everywhere

Fortis: A healthy spirit in a healthy body. This underlines the soldiers' physical fitness that enables them to fulfill their missions under any conditions.

Fidelis: Shows the soldiers' loyalty to their leaders and comrades, as well as their discipline and self-reliance in being able to think and act on their own.

A high level of training skill, together with vehicles possessing excellent mobility, make for a successful exercise. The highly mobile recce troops can train very realistically using the local terrain to best advantage. CVR(T) tracks have an average life of 5,000km (3,105 miles) of mixed road and cross-country travel. (Tim Mätzold)

A squadron Command Post (CP) is set up in a small wooded area, with two Sultans connected by a rear-mounted tent. A tactical map fitted to the vehicle's left side is used to brief the reconnaissance troop leaders. Usually the CP moves with the fighting troops and remains within easy access distance. (Daniel Nowak)

Advance to contact! A Scimitar of 1 JP makes contact with the enemy. After having quickly analyzed the situation, the commander has decided to attack. The 30mm Rarden L21A2 gun provides good penetration power against wheeled and lightly armored vehicles. The crew enjoys protection against rounds up to 7.62mm in caliber, or up to 14.5mm rounds to the vehicle's front. The coaxial turret-mounted 7.62mm L94A1 chain gun is an effective weapon against enemy infantry. (Tim Mätzold)

History of the Units

The first regiment to be established after the Belgian Revolution in September 1830 was the 1st *Chasseurs à Cheval*, which received its present name of "1 *Jagers te Paard*" in 1918. From that time until the ceasefire of 11 November 1918, 1 JP has been involved in all operational missions of the former cavalry division. After having used horses for more than one hundred years, the regiment became mechanized on 15 March 1938.

In 1952, the regiment was equipped with American materiel, but in 1973 the regiment began preparing to receive the new CVR(T) (Combat Vehicle Reconnaissance, Tracked) vehicles, which duly entered service in 1978. These units, which have been on continuous active duty since their establishment, were based at the following garrisons: Gent, Mechelen, Doornik, Bergen, Beverlo, Namur, Leuven, Brugge, and Leopoldsburg, which is their current garrison.

COMRECCE

In the days of COMRECCE (an independent reconnaissance unit operating at the 1st Belgian Corps level during the period of the Cold War), 1 JP was based at Arnsberg and Arolsen. 1 JP is currently the armored reconnaissance regiment of the 1st Mechanized Brigade based at Leopoldsburg. Until the end of the Cold War, COMRECCE was an independent mechanized reconnaissance unit attached to the HQ at Arolsen. It is noteworthy that this unit consisted of three armored reconnaissance regiments (1 JP, 2 JP and 4 ChCh), as well as one tank battalion (3.*Lansiers*), that were put under COMRECCE's command. This structure was discontinued with the end of 1st BE Corps.

Under ever-changing conditions, troop leaders deliberate about the best way to proceed to their objective as quickly as possible. This recce troop of 4 ChCh is moving in a reconnaissance corridor through the Norddeutsche Tiefebene (Northern German Bowl). This is only an exercise, but it still poses a challenge. (Tim Mätzold)

Passing through checkpoints was one task practiced during Exercise "Yellow Spring 2002" in Belgium, and this was supervised by umpires. The turret was traversed to the rear and a green flag was displayed, an internal sign for this exercise that allowed a vehicle to move on. (Tim Mätzold)

2 Jagers te Paard

2 JP was established in October 1830 as the "*Regiment Huzaren* Nr.8." They were involved in both World Wars as they defended their native country. The regiment was disbanded during WWII, their last shots being fired during the counterattack on Frezenberg on 27 May 1940. In 1952, the regiment was re-established and based in the German garrisons at Vogelsang, Arnsberg, Kassel, and Lüdenscheid. It was equipped with American materiel, including the famous M47 Patton tank.

In 1973, the unit began to change over to CVR(T)s. Later, the regiment changed to a tank regiment equipped with Leopard 1 main battle tanks in 1986. It served in this capacity until 1992. After the withdrawal of the 1st BE Corps from Germany, 2 JP was reduced to one squadron (*eskadron*) and moved to Altenrath near Köln, receiving a new role in the process. Since that time, 2 JP has been equipped with the SVB (*Slagveld Bewakings* Radar) battlefield surveillance radar fitted on M113 armored personnel carriers and Unimog L1350 trucks. Since June 2003, they have been located at Saive near Liège. In the future, 2 JP will receive another squadron in accordance with the new army structure known as "Landcomponent 2015."

This photo was taken just as deployment to the staging area was finishing. During Exercise "Green Sioux 2002," this Scimitar of 1 JP arrives at the exercise area near Gifhorn and begins its reconnaissance mission. The Iltis wheeled vehicle behind it belongs to the Sectie Communicatie, the detachment for press and public affairs personnel that always accompanies the troops. (Tim Mätzold)

4 ChCh

The 4th Regiment of *Chasseurs à Cheval* was established after the Belgian Army was engaged in combat operations around Antwerp in WWI. At the Yzer they were involved in trench warfare, having taken over an infantry role that they had not been prepared for. Nevertheless, they still accomplished their mission with pride. In 1918 they were deactivated, before being re-activated again in 1919. After having been based at the

The realistic sights presented to the camera make it easy to forget that this is just an exercise. This reconnaissance troop of 4 ChCh is impressive proof that Belgian recce units are highly trained. They have camouflaged their vehicles as well as possible with a combination of natural foliage and netting in an effort to catch their enemy off-guard. Even the lights are covered! (Tim Mätzold)

Field glasses, a map and their own human senses are the main tools of the recce team. In contrast to a modern reconnaissance vehicle, the Scimitar has neither a GPS system nor other kinds of technology, so the mission's success depends on the crew. (Tim Mätzold)

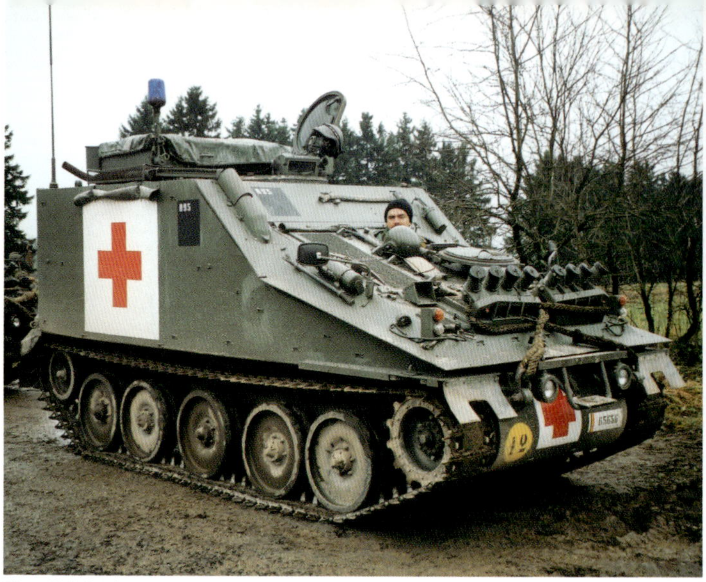

The FV104 Samaritan is used for medical evacuation under protection and in rough terrain. In accordance with the Geneva Convention, the vehicle is unarmed and is only fitted with 2x4 smoke dischargers for self-protection. The vehicle can be fitted either with four stretchers, five seats, or two stretchers on the right side and three seats on the left. The alloy-zinc-magnesium armor provides protection against 7.62mm rounds and grenade splinters. (Daniel Nowak)

Refueling in the field. In most cases, the fuel for the J60 6-cylinder petrol engine is transported in 20-liter (5.3 gallon) jerry cans. The fuel tank has a volume of 423 liters (112 gallons), which gives the vehicle an average road range of 550km (342 miles) based on a fuel consumption of about 76 liters/100km. The fuel consumption depends primarily on the terrain being traversed, and secondarily, the engine's condition. (Tim Mätzold)

4 ChCh used a new kind of camouflage consisting of Hessian strips and netting during Exercise "Yellow Integration" in September 2003. It is easy to attach, requires no natural material, and minimizes the vehicle's infrared signature. This Scimitar is sneaking around a small wood near Arlon, Belgium. (Tim Mätzold)

garrisons of Bergen and Krefeld, they were disbanded again in 1923, and finally re-activated during the mobilization of 1939.

On 1 February 1961, the 4th Regiment of *Chasseurs à Cheval* was established as the Armored Reconnaissance Regiment of the 1st BE Corps. They were based at *Kwartier Houthulst* at Werl, before finally moving to *Kwartier Reigersvliet* near Arnsberg in July 1964. The regiment served with COMRECCE until the end of the Cold War.

In accordance with the new army structure, the regiment was reduced to a squadron-sized force. On 3 June 1994, *Escadron* 4 ChCh took over the new role of Reconnaissance Squadron for the 1st Mechanized Division at their new garrison of Altenrath near Köln. On 1 January 2002, they became the Reconnaissance Squadron of the 7th Mechanized Brigade at Marche-en-Famenne, and then moved back to Saive, Belgium, in June 2003. On 1 January 2003, they returned to the regimental echelon, which prompted the establishment of an HQ Squadron and a second reconnaissance squadron with an HQ Troop. This process is not yet complete because their new vehicles have not yet been fielded. It should be mentioned that a non-active battalion with the title "5 *Chasseurs à Cheval*" existed from 1986 until the drawing down of the 1st BE Corps.

The Mission of Modern Armored Reconnaissance

The main tasks of the modern Belgian Armored Reconnaissance units are as follows:
Conducting reconnaissance missions
Securing areas, objectives and positions
Leading the battle
Capturing and holding objectives
These tasks are completed primarily by 1 JP and 4 ChCh, while 2 JP fulfils a different kind of reconnaissance role as dictated by their equipment. Their battlefield surveillance radar enables them to survey a wide area without the need for reconnaissance troops in the field.

Equipment and Structure

Equipment

The main vehicle that has been used by the Belgian Armored Reconnaissance force is the CVR(T) (Combat Vehicle Reconnaissance, Tracked), which represents a complete family of vehicles. This vehicle was developed by Alvis Ltd of Great Britain, and was manufactured under license at a Leyland plant in Belgium before it was put into widespread service. The Belgian Army ordered a total of 701 CVR(T) vehicles, including some 133 Scimitars and 153 Scorpions.

The anti-tank role was provided by the anti-tank platoon equipped with four FV103 Spartans armed with the MILAN anti-tank weapon, which can be fired either dismounted or from a vehicle-mounted launcher. The MILAN Spartan has a crew of five soldiers. The launcher of this Spartan, which is participating in Exercise "Yellow Bastard '99," is not fitted with the MILAN rocket container. (Clemens Niesner)

The MILAN, a product of the Franco-German consortium Euromissile, is a medium-range anti-tank weapon that has a maximum range of 2,000m (2,186 yards). The missile has a body diameter of 90mm (3.5 inches) and it weighs in at 6.65kg (15 lbs). The crew of this anti-tank Spartan is watching out for the blue forces taking the role of the enemy during Exercise "Yellow Mountain 2000" in northern Germany. (Clemens Niesner)

These reconnaissance vehicles have excellent mobility and provide protection against weapon rounds up to a caliber of 7.62mm (and up to 14.5mm in the frontal arc), as well as grenade splinters. The sophisticated alloy-magnesium-zinc armor allows the vehicle to keep its weight under nine tons. This results in a very low ground pressure of 0.338kg/cm², thus enabling the vehicle to negotiate wet and muddy ground. It can also traverse trenches up to 2.057m (6.75 feet) wide and vertical obstacles up to 0.5m (1.65 feet) high.

The Jaguar J60 No.1 Mk100B 6-cylinder petrol engine delivers 190bhp at 4,750rpm, propelling the vehicle to speeds of up to 80km/h (50mph) on roads! The semi-automatic TN15 gearbox, which was developed along the same lines as that used in the Chieftain MBT, includes seven forward and seven reverse gears, and it is relatively easy to use. It is shifted by a foot-operated gear lever so that the driver's hands always remain on the driving controls. With the gearbox in neutral, the vehicle can pivot quickly on its own axis – a noteworthy tactical advantage!

The basic reconnaissance vehicle is the FV107 Scimitar. It is fitted with a 30mm Rarden L21A2 gun and a 7.62mm coaxial L37A1 gun. With an ammunition storage capacity of 165 30mm rounds and 3,000 7.62mm rounds, the vehicle has considerable firepower. The gun can be elevated from –10° to +35°, and the turret can traverse 360°. All turret functions are controlled manually.

The crew consists of a driver, commander and gunner, who are all equipped with night vision devices. The commander is located on the left, and the gunner on the right. The commander has seven daytime periscopes and a roof-mounted day sight with a magnification of 1x and 10x in front of his hatch cover. The gunner has two daytime periscopes and a roof-mounted day sight with magnifications of 1x and 10x. A GEC-Marconi-Sensors passive night sight with a magnification of 5.8x and a low magnification of 1.6x is fitted to the right of the main armament.

Armored recce troops have to cover a wide area and travel long distances in a rather short period of time. To provide radio contact, it is necessary for the CP to move at the same speed at a safe distance from the fighting troops. This Sultan has apparently been attacked during a tactical move, so the commander is providing self-defense with his FN machine gun. (Clemens Niesner)

This Sultan of 1 JP's S2 section (the section responsible for military safety) has reached its assigned position, and the crew is now setting up the rear-mounted tent. The wooden flooring seen on the ground, which is used to keep muddy boots out of the vehicle, is usually fitted to the side of the vehicle during road marches. Note the Belgian flag on the antenna. (Clemens Niesner)

Belgian reconnaissance units can be deployed quickly by transport aircraft, a task that is practiced at least once a year. A total of two CVR(T)s can be carried in a C-130 Hercules. They are either unloaded or airdropped using a special airdrop pallet. Before the vehicles enter the aircraft, their weight has to be checked by mobile scales to avoid overloading. (Clemens Niesner)

Every vehicle is assigned its own position on the plane to ensure an equal load distribution. The wooden planks on the floor prevent damage to the aircraft by vehicle tracks. This airlift operation, which began at Bierset and ended at St. Hubert military airport, was part of Exercise "Yellow Integration 2003." 4 ChCh participated in the operation, which took place near Arlon in southern Belgium. (Clemens Niesner)

Structure

With the end of the Cold War, which prompted the disbandment of the 1st BE Corps, an armored reconnaissance regiment consisted of one Staff and Headquarters (HQ) squadron and three reconnaissance squadrons. It is noteworthy that 1 JP's regimental status remained unchanged following the Army's reorganization. It still has two active reconnaissance squadrons, while 4 ChCh was reduced to a squadron formation, before returning to a regimental status at the beginning of 2003.

Before unit realignments started taking place in the past few years, the Armored Reconnaissance Squadrons (1 JP and 4 ChCh) had the following structure:
Squadron HQ with two FV105 Sultans
Maintenance Section with one FV106 Samson and one FV103 Spartan Towbar
Medical Section with one FV104 Samaritan
Three Reconnaissance Platoons with seven FV107 Scimitars
One *Tirailleur* (skirmisher) Platoon for dismounted tasks with seven FV103 Spartans
One Anti-Tank Platoon with one FV103 Spartan as the lead vehicle and six FV103 Spartans equipped with the MILAN ATGW (Anti-Tank Guided Weapon).

Meanwhile, 2 JP was structured as follows:
Staff and HQ Squadron with *Cel Instructie* (Instruction Cell), which is responsible for the platoons' and section commander's radar calibration
Maintenance Group with Direct Support Element (DSE)
Two Troops with three Groups each consisting of Electronic Reconnaissance Tracked (M113) vehicles, as well as Electronic Reconnaissance Wheeled (Unimog L1350) vehicles. The latter is air-droppable.

The SVB (*Slagveld Bewakings* Radar) 2130A target surveillance radar is mast-mounted and can survey an area up to 30km (18.6 miles) wide, depending on the topography. It has a weight of 113kg (250 lbs). These specific objects can be identified out to the following distances:
Dismounted soldiers – 12km (7.5 miles)
Helicopters – 20km (12.4 miles)
Wheeled vehicles – 28km (17.4 miles)
Tracked vehicles – 33km (20.5 miles)
Detonations – 15km (9.3 miles)

In wartime conditions, 2 JP is supported by these other ground troops:
Artillery surveillance battery (Bij80A), Elsenborn
17 *Eskadron* Reconnaissance, Saive
LRRP (Special Reconnaissance)
Special Forces detachment

A typical armored reconnaissance troop usually consists of two Scimitars. In this case, during Exercise "Yellow Ambush" near St. Vith, Belgium, a Bombardier Iltis wheeled reconnaissance vehicle was the troop's first vehicle, followed by a Scimitar. Two crewmembers can rapidly carry out dismounted operations using their FN 7.62mm machine gun and L1A1 66mm anti-tank weapon, which has an average range of 150-300m (164-328 yards). (Clemens Niesner)

After having finished the first part of an exercise, 1 JP moves to a new staging area located some distance away. To avoid any problems, the battalion is guided along a prearranged route, which these men confirm by checking a road map. (Clemens Niesner)

Recent developments in Belgian Armored Reconnaissance Units

Although the Belgian reconnaissance units have had material and financial obstacles to negotiate (as has the Belgian Army generally), these problems have always been compensated for by the high motivation and professionalism of the troops. Even as the CVR(T) approached an average age of thirty years, it was still an effective reconnaissance vehicle, despite having been put through its paces on numerous occasions. The uncanny ability of the Belgian maintenance sections enabled the vehicles to keep going despite harsh treatment during missions and exercises. This is a big difference from CVR(T) users of other nations. However, the CVR(T) family has seen its last days, with their final live-fire exercise taking place at the Bergen-Hohne ranges in Germany in April 2004 before their withdrawal.

The Belgian Army has been quick to convert to its new structure under the auspices of "Landcomponent 2010". In the process, it has withdrawn almost all of its tracked vehicles, except for a few engineer tanks. When likely future missions (especially peacekeeping operations) were considered, it was decided a wheeled vehicle should take the place of the CVR(T) in the armored reconnaissance units. In recent years the Belgian Army has considered various wheeled vehicles to replace the CVR(T) family. Candidates have included the Dutch-German Fennek armored car, which is manufactured by Krauss-Maffei-Wegmann, and the British Scarab manufactured by Alvis. Another possibility was the Belgian armored vehicle manufactured by Sibmas.

In the end, the Belgian Army has decided to equip its armored reconnaissance units with the Pandur 6x6 and the Dingo. For the Dingo, there are currently several versions planned for use by the Belgium Army. These include:

Versie Troepentransport (Personnel Carrier) providing space for eight soldiers, including the BMS operator. It is armed with the ROSS (Remotely Operated Self-Defence Station) and a Battlefield Management System (BMS).

Versie Commandpost (Command Post Vehicle) with a crew of six soldiers, including one BMS operator and three operators for the Command, Control and Information System. It is fitted with ROSS.

Versie Ambulance (Ambulance Vehicle) provides space for a crew of three and five wounded soldiers. It is without armament but is fitted with BMS.

Versie Voorwaartse Waarnemer (Forward Observation Vehicle) with a crew of four, including two Forward Observers and a BMS operator.

A Belgian reconnaissance team in their element. Even though they know they are only on an exercise, the crew of this well-camouflaged Scimitar is doing the best job possible. To be able to get off the vehicle quickly, this commander prefers to ride on the vehicle's top deck, which is not an uncommon sight. (Clemens Niesner)

A Scimitar of 1 JP awaits contact with blue forces near Gifhorn. To make visual observation more difficult, the driver always tries to travel where the vehicle can blend in with the countryside. (Clemens Niesner)

In any case, the conversion to a new vehicle will be a big step for Belgian reconnaissance units. Their new vehicles will be fitted with such sophisticated technology as an inertial navigation system with a Global Positioning System (GPS). The main tools that the Belgian reconnaissance force has had to work with for a long time - map, glasses and their human senses – are about to be augmented with newer technologies!

After its recent reorganization, 1 *Jagers te Paard* currently has one Headquarters Squadron (*Eskadron Staf en Diensten*) and three reconnaissance squadrons (A, B and G *Eskadron*) equipped with the Bombardier Iltis light recce vehicle. It is noteworthy that A and B *Eskadron* are based at Leopoldsburg, while the newly established G *Eskadron* (which provides a parachutist capability) is based at Lombardsijde on the coast of Belgium. In 2007, two squadrons are supposed to be fully equipped with the Pandur 6x6 APC and one squadron with the Dingo 2 MPPV (Multi-Purpose Protected Vehicle).

The 4th Regiment of *Chasseurs à Cheval* amalgamated with 2 *Jagers te Paard* on 29 September 2004. It is now known as the 2/4 Regiment *Jagers te Paard*, and is comprised of:

- A *Eskadron* (reconnaissance squadron equipped with Iltis and Unimog vehicles [Pandur in the future] with three reconnaissance platoons).
- B *Eskadron* (Battlefield Surveillance and Target Acquisition, fitted with the EL/M 2130 NG/B Slagveldbewakingsradar SBV (Target Acquisition Radar), based on the M113 and Mercedes-Benz Unimog. Each platoon has three radar sections. It is noteworthy that all M113 vehicles will be replaced by Unimogs in the future. Additionally, a Direct Support Element (DSE) provides maintenance for the radar systems. The Belgian armored reconnaissance force is set remain an important part of the Army in the future, and its capabilities and strength will continue undiminished.

Acknowledgements

Many people have supported us in finishing this report. We especially want to say thank you to Major BEM Eric Kalajcik, Captain Pierard, Adjutant Chef Chapelle, LTC Hermann Geens, Adjutant Chef Vermaerke, and Filip van Reckem for their support.

The crews of these FV103 Spartans, belonging to the tirailleur peleton (skirmisher platoon) of B Eskadron of 1 JP, prepare to proceed. Exercise "Yellow Spring 2002" took place in Belgium, a fact revealed by the advertisement for "Jupiler," a famous Belgian beer! (Clemens Niesner)

This vehicle is not camouflaged for sitting beside a road! To support the recce troops, this highly camouflaged Unimog 1350L even used public roads during Exercise "Yellow Scrapper 2002," with its front windscreen completely covered. Two soldiers assigned to dismounted tasks sat in the back of the vehicle with the tailgate lowered, always ready to jump off. A third soldier provided air defense while standing on a wooden box located in the vehicle's cargo area. (Clemens Niesner)

The four-wheeled MAN 11.136 HA truck was produced only for the Belgian Army, and is still used in several variants. It is fitted with a MAN Savièm Model 797/06 6-cylinder diesel engine that delivers 136bhp. It also features a 5-speed gearbox with low gear ratios. (Clemens Niesner)

The MAN truck began service in 1975 and has an average age of more than twenty-five years, but they are still in service. Before its deployment on a KFOR mission, this truck from 1 JP was modified with special screen-protection grills to prevent damage from stone-throwing mobs. (Clemens Niesner)

The turret and the 30mm Rarden cannon on this Scimitar are operated manually by the gunner. Though an electric operating kit is available, it has never been fitted to any of these vehicles because they have never undergone a life-extension program. (Clemens Niesner)

The Unimog 1350L is the successor to its predecessor the Unimog U404S, which was withdrawn in the mid-1990s. This U1350L is a silent version, as may be seen by the closed half of the cooling grill. The front-mounted winch is driven by the power of the engine. (Clemens Niesner)

"Rats are released!" Belgian recce troops deploy to an FTX (Field Training Exercise) in northern Germany. These two Scimitars of B Eskadron, 1 JP, are seen moving at high speed through the village of Dorfmark at the beginning of Exercise "Yellow Mountain 2001." Once the reconnaissance troops are released, they drive to the staging areas as fast as their vehicles can take them. (Clemens Niesner)

The FV106 Samson, the CVR(T) family's recovery vehicle, is equipped with an engine-driven winch and an earth anchor. The winch has a length of 229m (250 yards) and a towing capability of 20.9 tons, using additional rollers. Using the winch's cable, the crew can erect a small crane. The earth anchor prevents the vehicle from having a larger rear door. (Clemens Niesner)

This Unimog U1350L of 2 JP also participated in Exercise "Yellow Spring 2002." It is fitted with the SVB battlefield surveillance radar. This particular vehicle was a troop trial, as is indicated by the lack of tactical signs. (Clemens Niesner)

A typical Belgian reconnaissance troop advances to contact. While on the move, the drivers use both sides of the road, thus leapfrogging and securing one another. Usually one vehicle moves ahead while the other is securing its advance. (Clemens Niesner)

This 4 ChCh Scimitar has a three-color camouflage scheme similar to NATO's standard camouflage, and is seen engaged in a troop trial. Generally, vehicles from 4 ChCh have a two-color camouflage scheme similar to that seen on British vehicles. (Clemens Niesner)

A Samaritan from 4 ChCh has reached its position to fulfill a medical evacuation task. In combat, the red cross is covered with a tarpaulin to prevent visual recognition. In most cases, the Unimog U1350L is used for this task due its superior comfort, equipment and road speed. (Clemens Niesner)

This FV103 Spartan Towbar is part of the maintenance section and is fitted with a wide range of spare parts and tools, along with a tow bar, to support the FV106 Samson. The right side of the flashing beacon is covered to prevent reflections from distracting or irritating the commander. (Clemens Niesner)

Securing objectives such as bridgeheads is a typical dismounted task, and it was practiced during Exercise "Yellow Spring 2002." The squad's vehicle remains in the background to minimize risk of observation. It follows at a safe distance, enabling it to pick up soldiers in case of an emergency. This soldier, who is equipped with an FN 7.62mm rifle, is carefully observing a riverbank. (Clemens Niesner)

A well-camouflaged Spartan of the fusilier peleton (rifle platoon) goes into a hide on private grounds during Exercise "Yellow Ambush 2002." It was the crew's task to prepare a trap in the village to engage approaching reconnaissance troops. (Clemens Niesner)

A sniper from 4 ChCh aims his Accuracy AW 7.62mm sniper rifle at a target. The special sniper outfit, which each man purchases for himself in most cases, gives each sharpshooter a unique look. (Clemens Niesner)

Two dismounted soldiers from the fusilier troop, equipped with an AW Accuracy 7.62mm rifle and a shoulder-launched LAW (Light Anti-Armor Weapon), watch for targets during Exercise "Yellow Bastard '99." (Clemens Niesner)

Belgian recce troops often work together with Dutch forces, as is seen in this photograph, which was taken during the KFOR mission "Belurokos VIII." A Dutch Chinook helicopter from the Tactische Helicopter Groep *(part of the* Koninklijke Luchtmacht*) has delivered spare parts to a maintenance point near Camp "Notting Hill". Note the fitter's blue work clothes, which are the usual combat attire. (4th Regiment of* Chasseurs à Cheval *via CPT Pierard)*

Setting up checkpoints was a primary task carried out during Operation "Belkos II", with two Scimitars used to guard each roadside. The dismounted soldier, who is wearing a ballistic vest, is responsible for controlling oncoming vehicles and persons. In the event of an emergency, the Scimitars will provide fire support. (4th Regiment of Chasseurs à Cheval *via CPT Pierard)*

Chilly weather conditions are no problem for the Jaguar J60 petrol engine, which was designed to be able to operate at temperatures from –30° to +50°C (-22° to +122°F). Troops involved in KFOR missions must often operate in wintry weather. (4th Regiment of Chasseurs *á* Cheval *via CPT Pierard)*

The way of camouflaging vehicles has not changed for the Belgian armored reconnaissance units - a mix of a camouflage net combined with Hessian strips makes this vehicle blend in with the environment. The roof has been completely removed, and the windscreen has been folded down and protected by a cover. This way of driving is not always comfortable for the crew, plus space is always limited. (Tim Mätzold)

A reconnaissance team of 1 Jages te Paard makes contact with the enemy on Exercise "Quick Response 2006" near Dinant, Belgium. The driver keeps in radio contact whilst the vehicle commander operates the FN MAG 7.62mm machine gun. (Tim Mätzold)

The Anti-Tank Peleton (Anti-Tank Platoon) that was at one time equipped with the Spartan CVR(T), is currently equipped with the Iltis and the man-portable MILAN launcher. Compared to the Spartan, the crews now have less comfort in their vehicles. They have no rear compartment to provide comfortable housing when out in the field, and instead have to set up tents. This photograph was taken on Exercise "Fatextel 2006" in France. (Clemens Niesner)

A reconnaissance squadron currently comprises three reconnaissance peletons (platoons) with seven Bombardier Iltis (4x4) recce vehicles each, plus one Peleton Voltigeur with Unimog trucks (for carrying out infantry tasks). The Peleton Voltigeur only exists within A and B Eskadron.

This Iltis belongs to G Eskadron, 1 Jagers te Paard. These vehicles can be identified by their wider tires, this feature being due to the airdrop capability of that squadron. Note the gun mount for fitting a top-mounted machine gun. (Tim Mätzold)

Maps and binoculars are still the most important tools of the Belgian reconnaissance units. The only "Optronic" is a handheld night vision device to support operations in the dark. This recce team has just crossed the "Canal to Beverlo" near Leopoldsburg, during Exercise "Quick Response 2006" in Belgium, and awaits enemy contact shortly behind the crossing site. At the time of writing it is believed that the Pandur will not be fitted with any special reconnaissance optronics. (Tim Mätzold)

This squadron headquarters of A Eskadron, 1 Jagers te Paard, is now transported in Unimogs instead of Sultan CVR(T)s. The photograph was taken at the village of Salet in southern Belgium on Exercise "Quick Response 2006". (Tim Mätzold)

This picture shows the M113 with the Slagveld Bewakings Radar SBV in working position; the radar unit is mast-mounted and can be erected above the vehicle. These vehicles will be replaced by the Unimog. (Clemens Niesner)

The Unimog 1350L will be the main vehicle of B Eskadron, 2 Jagers te Paard, and will carry the Slagveldbewakingsradar SBV (Target Acquisition Radar) EL/M 2130 NG/B. This system provides deep surveillance and reconnaissance of the battlefield using sophisticated technology, and so supports the formation reconnaissance carried out by the recce squadrons in their Iltis vehicles. (Clemens Niesner)

The Pandur 6x6 wheeled APC will be the future reconnaissance vehicle of Belgian armored reconnaissance units. The 13.5-ton vehicle is powered by a Steyr WD612 diesel engine that develops 265bhp. A maximum speed of 110km/h can be achieved, while the vehicle has an average range of 650km. (Tim Mätzold)

According to the new structure, the Dingo 2 MPPV (Multi-Purpose Protected Vehicle) will equip one reconnaissance squadron of 1 Jagers te Paard. There are currently five versions planned for use by the Belgian Army. The Dingo 2 is manufactured by Krauss-Maffei-Wegmann in Germany. (Robert Zwaart)